PINKI

设计15年

PINKI DESIGN IN THE COURSE OF 15 YEARS

刘卫军 编著

MANAGE TO REVIVE CLASSICAL HUMANISTIC AESTHETICS BY DESIGNING

江苏凤凰科学技术出版社

图书在版编目（CIP）数据

PINKI设计15年 / 刘卫军编著. -- 南京：江苏凤凰科学技术出版社，2017.6
 ISBN 978-7-5537-7175-5

 I. ①P… Ⅱ. ①刘… Ⅲ. ①室内装饰设计－作品集－中国－现代 Ⅳ. ①TU238.2

中国版本图书馆CIP数据核字(2016)第215407号

PINKI设计15年

编　　　著	刘卫军
项 目 策 划	凤凰空间
责 任 编 辑	刘屹立　赵　研
特 约 编 辑	杜玉华

出 版 发 行	凤凰出版传媒股份有限公司
出版社地址	南京市湖南路1号A楼，邮编：210009
出版社网址	http://www.pspress.cn
总　经　销	天津凤凰空间文化传媒有限公司
总经销网址	http://www.ifengspace.cn
印　　　刷	上海利丰雅高印刷有限公司

开　　　本	965 mm×1270 mm　1/16
印　　　张	27.25
字　　　数	212 000
版　　　次	2017年6月第1版
印　　　次	2017年6月第1次印刷

标 准 书 号	ISBN 978-7-5537-7175-5
定　　　价	398.00元

图书如有印装质量问题，可随时向销售部调换（电话：022-87893668）。

序言一

不刻意，不强求，做自己，这大概是刘卫军与身俱来的特质。他将这种独一的特质融入自己的每一件设计作品当中，没有规矩，没有束缚，没有特定的方向；他用对美的偏执，创造出各种意境生活与故事。

—— 设计巨蛋

PREFACE ONE

Danfu Lau is born to be himself regardless of attempted intention. With his own unique character, he accomplishes every design project with no rules, constraints and set purposes. Guided by his firm belief in beauty, he has created various artistic living styles and stories.

— by Design Dome

序言二

每个梦想有门可入

记得丙申年春时，刘兄来电谈起品伊15周年出书之事，电话里听得到那份信任，我欣然接受。继《灵感塑造空间》的出版6年过去了，品伊在刘兄的带领下走过了露往霜来十五载，新书出版也是他托物寓感恰好的事。受人所托，忠人之事，我也寄望新书的顺利出版，可以让更多朋友看到今日之品伊，依然不负前行志始的品伊人。

在我看来做书尚可，写序就汗颜了，希望自己的寥寥闲叙别耽误了读者的兴致。我在浏览品伊网站时，偶然看到品伊企业招聘季的广告主题："每个梦想有门可入，无需偶遇，青春回甘"。多少年，品伊人用这样的文字，不断叙述着自我对坚守、对品行、对企业精神的理解和要求。作为品伊的老朋友，我深知刘兄百舍重茧的不易，也看到了"品伊式的梦想"，日日行，不怕千万里；常常做，不怕千万事。我借用"每个梦想有门可入"这个招聘广告作为标题，着实是我真切对品伊人所做所行的感受。

从创办品伊到今天的品伊国际创意美学院，从设计师到职业导师，从授业再到育人，刘兄身份的切换总是惹人注目。有一次与他在上海会面，聊到先行者的忧患，看得出刘兄目注心营的背后生来有颗"不安分的心"，也因这"不安分的心"他收获了独有的人生剧本。一晃许多年过去了，我们会感怀一个创业者的守定，更会为他的致知力行而点赞。用刘兄自己话说"只要路对，就不怕路远"，我相信品伊还有很远的路要走，由衷希望刘兄达成所愿，行疆大道。

落笔丁酉立夏

镗铁传播创始人　张斌

Directory
目录

008
Connotation Of Brand
品牌诠释

010
Brief Introduction Of Pinki Design
PINKI DESIGN 简介

012
Brief Introduction Of Danfu Lau
刘卫军简介

016
Design Theory
设计理论

022
Great Artist
大艺术家

042
Great Artist - Form
大艺术家之源

056
Great Artist - Source
大艺术家之形

070
Great Artist - Chamber
大艺术家之厢

082
Great Artist - Autumn
大艺术家之繁秋

100
Great Artist - Spring Rhapsody
大艺术家之春光赋

116
Great Artist - Perch
大艺术家之栖

124
Time Outside
遇外时光

144
The Contemporary Esthetician
大时代美学家

168
Beautiful World of Winter Fairy Tale
冬日童话里的美丽世界

188
Koi and Begonia
大鱼海棠

198
Diffused Time
唯漫时光

206
Azure
青蓝幽境

228
Connoisseur
鉴藏

236
Uncharted
圆中秘境

246
Corridor of Time
时光走廊

252
Oriental Charm
韵魅东方

262
Hao Sheng Yin Yi
豪笙印溢

270
Drunken Garden of the Peaches
醉慕桃源

280
Sense - Time Traveler
感·时光旅者

290
Baroque Love Song
巴洛克恋曲

302
Song of Flowers and Full Moon
花好月圆曲

314
Waves of Flower
花漾漫波

328
Water Garden
水榭听香

338
Splendor of Andorra
安道拉的璀璨

350
California Cottage
加州馨居

360
Morning Love Song
晨光恋曲

372
Great Architect Spectacular Time
大建筑家·曼妙时光

382
Royal Incense
御品倾香

392
Contemporary New Home Your Neighborhood
时代新居与你同邻

414
Legend of Love and Dream
俊逸山房

424
LIST OF AWARDS
奖项列表

431
LIST OF CLIENTS
客户列表

432
Team culture
团队文化

品牌诠释

Promotion 提升
Imagination 想象
Notion   见解
Keenness 敏锐
Ingenuity 创造

品，作名词使用，取品行之意；作动词使用，取玩味之意。伊，作代词使用，取万物之意，衍生创造与包容之意。品伊，包含对立与统一的哲学，用乐观的心态包容万物。

多角度多角色地辨别与提升价值，兼具创造性与平衡性。倡导在创意设计、文化美学与艺术商业中寻求共鸣与平衡。

品牌理念
创意，实现梦想

服务理念
以心为·由心造

品牌形象
向日葵

品牌文化
爱与包容

文化核心
创造 梦想 行动

品伊人形象
创意梦想家与行动派

CONNOTATION OF BRAND

品(pǐn), a Chinese character, means either "behavior" if used as a noun or "appreciation" if used as a verb. 伊 (yī), as a pronoun, means "everything in the world", and furthermore means "inclusiveness".

PINKI's brand connation is related to the law of the unity of opposites. With a positive attitude, PINKI is bound to be an inclusive brand. PINKI brand gets promoted in different aspects by playing different roles, and meantime it's of creativity and balance. In addition, PINKI pursues consensus and balance among creative designs, cultural aesthetics and artistic business.

BRAND CONCEPT
Creativity makes dreams come true.

BRAND IMAGE/ LOGO
Sunflower

CORE CORPORATE CULTURE
Creativity Dream Action

SERVICE CONCEPT
From the heart To the heart. Follow your heart.

CORPORATE CULTURE
Love and inclusiveness

MEMBERS OF PINKI
Creative dreamers and actions

PINKI DESIGN 简介

PINKI DESIGN是PINKI品伊国际创意持续稳健发展的顶级地产设计知名品牌。2000年创立于国际设计之都深圳，由知名设计师刘卫军先生主持创建并担任事务所首席创意总监。历经17年地产设计创意实践，从为客户提供高端别墅、酒店、会所创意设计方案，延伸至更为精准的客制化服务。

公司团队本着"以心为，由心造"的服务理念，将"快乐设计，快乐生活"的生活观融入创作中，屡获国内外权威大奖120多项。先后被评为：IAID全国最具影响力设计机构、深圳最佳室内设计公司、地产界最具合作价值设计机构、中国最佳住宅室内设计十强企业、中国最具价值的室内设计十强企业、中国十大最具潜力设计机构、大中华地区最具影响力设计机构、国际设计之都深圳企业文化建设优秀单位。

荣誉见证使命与责任，团队坚持"创意乃设计之本"，以新感官主义·情感美学定位设计体系、高专业度、高服务品质、跨域品牌资源整合相结合的超预期客户服务，全方位地用文化艺术+美学创意为产品提升商业与艺术价值。PINKI DEISGN认为，没有什么比精神更值得保留，创新创意，创造价值。

BRIEF INTRODUCTION OF PINKI DESIGN

PINKI DESIGN is a top real estate design brand constantly and steadily developed by PINKI Group. It was founded in Shenzhen, the capital of international design in 2000 by a famous designer Mr. Danfu Lau, the Chief Creative Officer. From providing creative design proposals for high-end villas, hotels and clubs to more précised customizing services, it has had experiences of 17 years in real estates' creative design practices.

With the company's serving concept of "From the heart, To the heart", it has integrated the lifestyle of "Happy design, Happy life" into its creation. The company was awarded with top authorized awards for more than 120 times at home and abroad, such as "IAID The Most Influential Design Agency In China", "The Best Interior Designer In Shenzhen", "Best Cooperate Value Design Agency In Real Estate", "Chinese Top 10 Residential Interior Designer", "Chinese Top 10 Interior Design Company with Best Value", "Chinese Top 10 Design Agency with Potential", "The Most Influential Design Agency in Greater China" and "Award of Excellent Cooperate Culture in Shenzhen, the Capital of International Design".

Those honors have witnessed the company's missions and responsibilities. Sticking to the concept of "Creativity is the root of design", the team provides exceeding-expectation customer services with designing system based on Neo-Sensualism Emotional Aesthetics, high specialization, outstanding service quality and cross-brand resource integration, using arts and aesthetics creativity on all aspects to enhance products' commercial and artistic value. PINKI DEISGN believes, "When it comes to bring about creativity and value, nothing is more valuable than spirit."

PINKI DESIGN

刘卫军
DANFU LAU

IARI 美国国际注册与认证 高级室内建筑师
IARI（International Accreditation
and Registration Institute）

刘卫军简介
BRIEF INTRODUCTION OF DANFU LAU

艺术家，设计师，PINKI品伊国际创意品牌创始人，2009年中国时代新闻人物，被誉为"空间魔术师""中国最具商业价值创造设计师"，2013年被授予中国室内装饰协会成立25周年"中国室内设计教育贡献奖"。

首创新感官主义·情感美学设计理论体系，为首登《亚洲新闻人物》的中国设计师，IARI美国（国际认证及注册协会）注册高级设计师，2015年入选（纽约）联合国70+华人当代艺术·创意设计成就展，全球创意设计大赛执行委员会副主席，国际设计之都（深圳）专家库11位成员之一，CIID学会第一个代表中国设计界赴韩国首尔担任"第四届亚洲室内设计联合会年会暨国际室内设计学术交流会"演讲人，CIID学会第一个亚洲室内设计论文奖获得者，获Asia-Pacific Interior Design Awards（亚太室内设计大赛）金、银、铜奖，获IHFO中外酒店白金奖2013年度最佳创意设计白金奖，获2015CWDA首届国际橱窗设计人赛特别贡献奖，获2015澳门"金莲花"杯国际设计大师邀请赛 金奖，获2016台湾TAKAO室内设计大赏商业空间金奖，获APDC赞颂·创造力2015/2016亚太室内设计精英邀请赛银奖，获2016法国双面神INNODESIGN PRIZE国际设计大奖银奖，2009年投资创建支持政府城镇化项目"中国城镇居住环境新理念观摩示范基地（中国首例）"，访学世界20多个国家地区，作品囊括超过120项国际和亚太区设计及企业、个人奖项。

Mr. Danfu Lau was the founder of PINKI design brand, and one of China Times News Figures in 2009. He is known as the "Space magician", and was also honored with "Chinese Designer with Most Business Value" and "Award of Contribution in Interior Design" by China National Interior Decoration Association at its 25th founding anniversary in 2013.

He initiated the theory system of Neo-Sensualism Emotional Aesthetics, becoming the first Chinese designer on "Asian News Maker". He is also a registered senior designer at IARI (International Accreditation and Registration Institute), and was chosen at The United Nations 70+ Chinese Contemporary Art & Originality Design Exhibition (New York). He is also the vice chairman at Global Creative Design Competition's Executive Committee, and one of the eleven expert tank members in the capital of international design (Shenzhen). He was the first person to represent Chinese designers at CIID and presented at Asian Interior Design Institute Association Annual Meeting & International Interior Design Workshop in Seoul, South Korea as well as the first one to win the Award of Asian Interior Design Paper at CIID. He was awarded with gold, silver and bronze prizes at Asia-Pacific Interior Design Awards, Platinum Award of IHFO China and Oversea Hotels, Platinum Award of the Best Creative Design in 2013, Award of Special Contribution at 2015 CWDA 1st International Window Display Contest, Golden Prize of 2015 Macao "Golden Lotus" International Design Invitational Tournament, Golden Prize of Business Space at 2016 Taiwan TAKAO Interior Design Awards, Silver Prize of APDC CelebratingCreativity 15/16 Asia-Pacific Interior Design Awards for Elite, and Silver Prize of 2016 France INNODESIGN PRIZE. He has invested and supported the government's urbanization project "Demo Base of China Urban Living Environment New Theory (First in China)" in 2009. He has travelled to more than 20 countries and regions as a visiting scholar, and his works have covered more

PINKI DESIGN

像蚂蚁一样工作，像蝴蝶一样生活。

Danfu Lau
刘卫军
Founder & Creative Director
创始人 & 创意总监

With limited energy to create unlimited wonderful!
用有限的精力创造无限的精彩！

Fullcy Zhong
钟文萍
Founder & Executive Director
创始人 & 执行董事

Good learning, Good intentions !
好好学习，天天向善。

Shirly Lee
李莎莉
Art Director
艺术总监

Life is not an end, but a journey.
生活不是目的，是旅程。

Ivy Luang
梁毅飞
Administrative Director
行政总监

Working like ants，Living like butterfly.
像蚂蚁一样工作，像蝴蝶一样生活。

Maggie Mei
梅玫

There is no such thing as failure, unless you stop trying.
没有所谓的失败,除非你不再尝试。

Tony Chan
陈春龙

Design is a kind of the perception and create of the world, is an attention of the mind。
设计是对外部世界的感应、创造,是心灵的观照。

Simon Luo
罗胜文

Emotional thinking, rational transformation and implementation。
感性的思维,需要理性的转化与实现。

Aimee Zhang
张慧超

Always keep the heart of fear, of sunshine, to beauty, to pain.
要始终保持敬畏之心,对阳光、对美、对痛楚。

Junhao Li
黎俊浩

Let the inspiration free to create the dreaming space。
让灵感自由释放,创造一个心灵渴望的空间。

Peter Fan
范彬

Brand is a kind of design about the feelngs of people, is defined and redefined。
品牌是一种关于人的感受的设计,是定义与重新定义。

设计理论
2005 年

2005年 提出主题式空间设计方法

产生的时代背景：在中国经济蓬勃发展的今天，室内设计专业领域也随着时代的变迁而有了极度快速的变化，同时也涌现出众多非常有自我个性的设计新锐，重新界定了多元化专业领域的视觉审美定义，极度地促进了室内设计行业的发展，增强了专业理念的多样性与时代个性。主题式设计方法正是因应了时代发展的需求，为空间设计创造富有艺术、商业、生活体验的价值。

主题式空间设计是什么？是一种能充分发挥自我认知、认识、认可的文化思想定义及开启设计灵感线索的依据，从而使作品实现多样性创作，避免雷同，强化了作品的创意设计价值，使作品的表现更具唯一及独特的商业价值。它区别于所有设计风格，又包含所有设计风格。

我们在设计每个项目的过程中都具有不同的情感诉求，也不知不觉地融汇了商业需求与设计思想表达的愿景，促使了一种新形式设计方法的形成。

主题式设计方法，也是一种行为思想与参照物的表现，具有很强的模拟性与想象性，更容易催发空间的感染力与商业诉求，以求达成目的性的设计意愿。

衍生创意设计方法
意，情理之中，意料之外；
乱，多元自由，无章主义；
情，情绪情感，情调情怀；
迷，异域风情，神秘未知。

DESIGN THEORY
In 2005

IN 2005, WE PROPOSED THE THEMED SPACE DESIGN METHOD.

Era background: With China's economy booming and flourishing today, the field of interior design has also evolved rapidly with the fast changing of the times. Meanwhile, a number of individualized and promising young designers have sprung up to redefine the diversified professional visual aesthetic definition, which has dramatically promoted the development of interior design industry, and enhanced the diversity of professional concepts and era characters. The themed space design method suits the needs of the society, providing artistic and commercial value for space design.

What is the themed space design? It is a kind of cultural definition which gives full play to the self-awareness, understanding and recognition, and also a basis of sparking the design inspiration.
So that it not only realized the manifold art creation, avoidance of similarity, but also improve the works' creative design value, reinvest the work with unique and special commercial value. It is different from all the design style, and at the meantime contains them.

We have different emotional demands in the process of designing each project, and integrating the business needs and design ideas unconsciously, which facilitating the formation of a new design method.

The themed design method is also unity of behavior thought and reference object, which is full of simulation and imagination. It is easier to simulate space attractions and create commercial demands in order to accomplish the design aim.

Method of derivative creative design
Meaning: reasonable but unexpected;
Diversification: various, free and not restricted;
Emotion: sincere sentiment and artistic;
Charm: exotic and mysterious .

设计理论
2013 年

2013年 提出新感官主义情感美学设计理念

基于形式追随功能的设计理念与对主题式空间设计方法的优化与革新，主张以情感美学创造艺术空间，其美学取向分为人本取向与文本取向两种，人本取向以人的感性自我为主体，表现为非理性和先验化的审美向度；文本取向以文本为主体，表现为形式化与语言化的审美向度。

深入研究人与物、情与理的关系，以空间实体存在作为空间使用者的情感与室内装饰品相统一的共同基础，将人的五感体验注入空间创作，将创意表现形式特征加以理性化，使空间的艺术性与实用性达到平衡。

DESIGN THEORY
In 2013

IN 2013, WE PUT FORWARD THE NEW SENSUAL AND EMOTIONAL AESTHETIC DESIGN CONCEPT.

Based on the design concept of the form following the function and the optimization and innovation of the themed space design method, it advocates creating artistic space by emotional aesthetics. The aesthetic orientation is divided into humanistic orientation and text orientation. The former is based on human sensibility, whose feature is irrational and transcendental aesthetic dimension; the latter regards text as the main body, whose feature is the formalized and linguistic aesthetic dimension.

After having lucubrated the relationship between people and things, emotion and reason, we have used space entities as the conjunct foundation of the user's emotion and interior decorations. Integrating the human faculties and senses into the space creation, rationalizing creative expression features, in this way can we achieve the balance between space decoration and practicability.

PINKI DESIGN

设计理论
2017 年

2017年 用设计复兴人文美学经典

大艺术家
重新定义新东方美学设计方式，让艺术的生活空间交融于设计美学之中，探寻另一种新形式的可能，适用于当代审美的认知，回归时代设计的初心，建立多面体的新的生命载体。

大生活家
以传统与当代的人文转化、以人与自然的情感融合、以新的感官体验来开启一场人文美学复兴之旅。

大建筑家
以点、线、面的方式，构建出任何形式组合而成的立体空间，传达理性与感性的设计美学效应，用最具时代感的语言传递。摒弃模式化的设计方式，创造一幅与时俱进的、生动平实的、流动的生活风景画。

大时代美学家
崇尚自然，让心灵回归本源，提供一种心随自然的美的享受、一场浸淫人文的华丽体验，用匠心传承的精神，去传导空间的细节，营造一个富有品质与时代感的生命体验空间。

DESIGN THEORY
In 2017

IN 2017, WE MANAGE TO REVIVE CLASSICAL HUMANISTIC AESTHETICS BY DESIGNING.

The great artist
We redefine the new oriental aesthetic design method, integrating the artistic living space in design aesthetics to explore the possibility of another new design form, which applies to contemporary aesthetic cognition, return to the design essence of the times, and establish a new polygonal vehicle of life.

Living artist
We will lead you to have a humanistic aesthetic renaissance trip with new sensory experience, which involves humanistic transformation and the emotional integration of human and natural.

The great architect
We can construct various stereoscopic space by point, line to plane, to convey the rational and emotional design aesthetics with the most contemporary language. Abandoning the stereotype of design, we try to create vivid and splendid life landscape painting.

The contemporary esthetician
We advocate staying true to nature, which enables you to attribute your emotion to its source. And then we start a gorgeous journey and get immersed with humanities and enjoyment. We inherit culture with ingenuity and pay more attention to details to create a elegant and fashionable house.

GREAT

ARTIST

PINKI DESIGN

Topic	Great Artist
Project	PINKI Creative Art Design Institute
Customer	PINKI International Creative Group
Unit	Office
Size	1480 square meters
Location	Shenzhen
Created by	PINKI DESIGN US IARI LiuWeiJun Design Firm
Interior Decoration and Implementation	TATS Artists Interior Decoration and Implementation
Primary Materials	Marble, wooden finishes, ceramic tiles, wood flooring

主题名称	大艺术家
项目名称	PINKI品伊创意艺术设计研究院
客户名称	PINKI品伊国际创意集团
户　　型	办公室
设计面积	1480平方米
项目地点	中国深圳
创意出品	美国IARI刘卫军设计事务所
软装设计与实现	TATS大艺术家软装设计与实现
主要材料	大理石、木饰面、瓷砖、木地板

GREAT ARTIST

大艺术家

PINKI品伊国际创意集团　中国 深圳

The space of creativity and design, in the magnificent sea of thoughts, at this very moment, we must learn the courageous actions that defy death, and swimming toward the desolate island of absolute silence while holding onto the lifesaver.
— Danfu Lau

创造设计的空间
在波澜壮阔的思海中
此刻瞬间
要学会勇敢视死般的行为
抱着救生圈游向绝对苍凉寂静的孤岛
——刘卫军

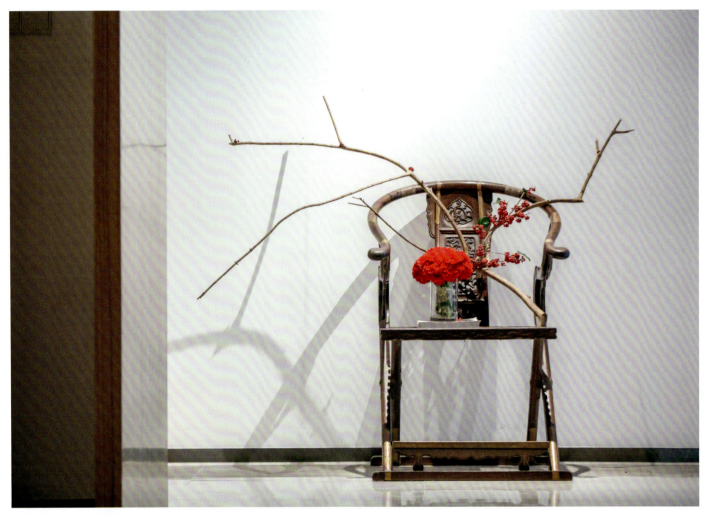

PINKI DESIGN

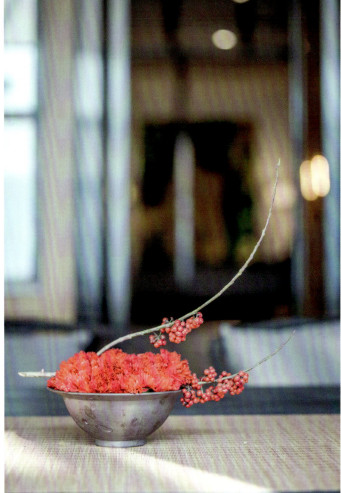

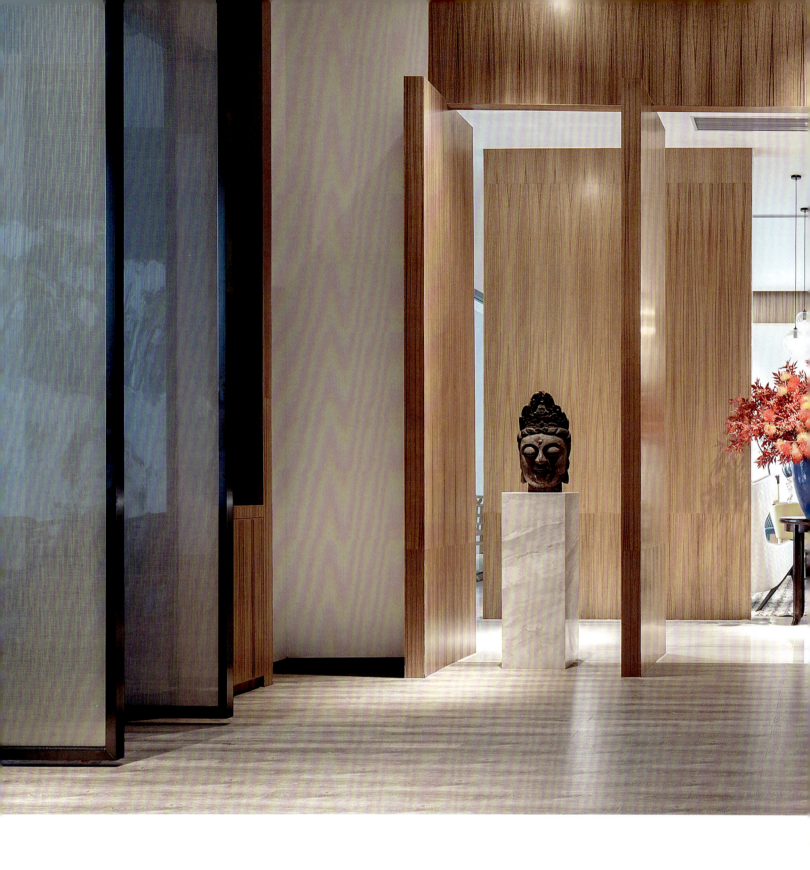

 GREAT AR

ST - SOURCE

Topic Great Artist - Source
Customer China Jinmao
Unit Unit D11
Size 613 square meters
Location Guangzhou
Created by PINKI DESIGN
Interior Decoration and Implementation TATS Artists Interior Decoration and Implementation
Primary Materials Marble, wooden finishes, tapestry, art glass, metal
Awards 2016 Taiwan TAKAO Interior Design Awards Commercial Space Category Gold Award, 2016 Macao Design Exhibition "Golden Lotus" Cup International Design Master Invitational Gold Award, 2016 ID + G "Golden Creative Award" International Space Design Awards Real Estate Space/Display Art Silver Award, 2016 Eighth China Lighting Applications Design Competition Silver Award.

主题名称 大艺术家 · 源
项目名称 广州南沙金茂湾商墅
客户名称 中国金茂
户　　型 D11户型
设计面积 613平方米
项目地点 中国广州
创意出品 PINKI DESIGN品伊设计机构
软装设计与实现 TATS 大艺术家软装设计与实现
主要材料 大理石、木饰面、墙布、艺术玻璃、金属
荣获奖项 2016台湾TAKAO室内设计大赏商业空间类金奖、2016澳门设计联展"金莲花"杯国际设计大师邀请赛金奖、2016 ID+G"金创意奖"国际空间设计大奖地产空间类/陈列艺术类银奖、2016第八届中国照明应用设计大赛银奖。

ARTIST THINKING

GREAT ARTIST - SOURCE

大艺术家·源

中国金茂集团　中国 广州

It is very difficult to do art, but even more difficult than art is life, finding a way to live life with an attitude of art. Breaking the traditional oriental aesthetic sense, opening up artistic design concepts, creating the shape and meaning of three-dimensional space…
— Danfu Lau

做艺术很难
比艺术更难的是生活
如何用一种艺术的态度去生活
打破东方传统美学意境
开启艺术设计意念
创造立体空间的形意
——刘卫军

GREAT AR

IST - FROM

Topic Great Artist - Form
Project Guangzhou Nansha Jinmao Bay Villa
Customer China Jinmao
Unit Unit D11
Size 613 square meters
Location Guangzhou
Created by PINKI DESIGN
Interior Decoration and Implementation TATS Artists Interior Decoration and Implementation
Primary Materials Marble, wooden finishes, tapestry, art glass, metal
Awards 2016 Macau International Design Exhibition "Golden Lotus" Cup International Design Master Invitational Gold Award, 2016 ID+G "Golden Creative Award" International Space Design Awards Exhibition Art/Real Estate Space Bronze Award, APDC Praise/Creativity 15/16 Asia Pacific Interior Design Elite Invitational Award, "2016 China Interior Design Yearbook" selected collection.

主题名称 大艺术家·形
项目名称 广州南沙金茂湾商墅
客户名称 中国金茂
户　　型 D11户型
设计面积 613平方米
项目地点 中国广州
创意出品 PINKI DESIGN品伊设计机构
软装设计与实现 TATS大艺术家软装设计与实现
主要材料 大理石、木饰面、墙布、艺术玻璃、金属
荣获奖项 2016澳门国际设计联展"金莲花"杯国际设计大师邀请赛金奖、2016ID+G金创意奖国际空间设计大奖陈列艺术类/地产空间类铜奖、APDC赞颂·创造力2015/2016亚太室内设计精英邀请赛佳作奖、《2016中国室内设计年鉴》入选收藏。

ARTIST THINKING

GREAT ARTIST - FORM

大艺术家·形

中国金茂集团　中国 广州

Redefining the new oriental design approach, infuse the artistic life into aesthetic design, explore the possibility of another form, return to the beginning of design, establish a new carrier of life for the hexahedron space.
—— Danfu Lau

重新定义新东方设计方式
让艺术生活交融于设计美学之中
探寻另一种形式的可能
回归设计的初心
建立空间六面体新的生命载体
——刘卫军

GREAT ARTI

T - CHAMBER

Topic	Great Artist - Chamber
Project	Guangzhou Zhujiang Jinmao Mansion Model Suite
Customer	China Jinmao
Unit	Floor 5, unit D-2
Size	142 square meters
Location	Guangzhou
Created by	PINKI DESIGN
Interior Decoration and Implementation	TATS Artists Interior Decoration and Implementation
Primary Materials	Marble, wooden finishes, tapestry, art glass, metal
Project Awards	2016 Taiwan TAKAO Interior Design Awards Single-story Residential Space Category Award, 2016 Meiju Award China's Most Beautiful Model Suite

主题名称	大艺术家·厢
项目名称	广州珠江金茂府样板房
客户名称	中国金茂
户　　型	5号楼D-2户型
设计面积	142平方米
项目地点	中国广州
创意出品	PINKI DESIGN品伊设计机构
软装设计与实现	TATS大艺术家软装设计与实现
主要材料	大理石、木饰面、墙布、艺术玻璃、金属
项目获奖	2016台湾TAKAO室内设计大赏单层住宅空间类优选奖；2016美居奖中国最美样板间。

ARTIST THINKING

GREAT ARTIST CHAMBER

天艺木家 · 厢

中国金茂集团　中国 广州

One flower is a paradise, a blade of grass is a world, one thought is one cleansing, one room is the land of the pure. Come, I am waiting for you right here…
— Danfu Lau

一花一天堂
一草一世界
一念一清净
一厢一净土
你来，我就在这里等你
……
——刘卫军

GREAT ART

ST - AUTUMN

PINKI DESIGN

Topic	Great Artist - Autumn	主题名称	大艺术家 · 繁秋
Project	Foshan Lake Ludao Project Villa Model Suite	项目名称	佛山绿岛湖项目别墅样板间
Customer	China Jinmao	客户名称	中国金茂
Unit	Unit 250	户　　型	250户型
Size	530 square meters	设计面积	530平方米
Location	Foshan	项目地点	中国佛山
Created by	PINKI DESIGN	创意出品	PINKI DESIGN品伊设计机构
Interior Decoration	TATS Artists Interior Decoration and Implementation	软装设计	TATS大艺术家软装设计与实现
Primary Materials	Marble, wooden finishes, tapestry, metal, art glass, wood flooring	主要材料	大理石、木饰面、墙布、金属、艺术玻璃、木地板

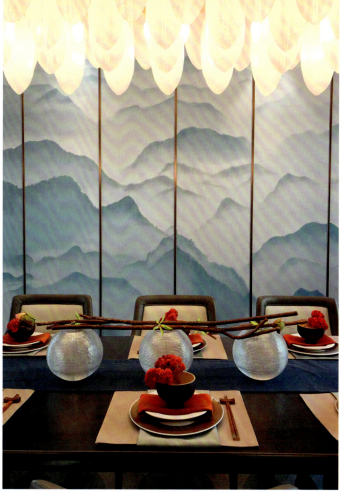

GREAT

SPRING

ARTIST

HAPSODY

PINKI DESIGN

Topic Great Artist - Spring Rhapsody
Project Guangzhou Tianhe Jinmao Suqare Clubhouse Sales Center
Customer China Jinmao
Unit Clubhouse
Size 950 square meters
Location Guangzhou
Created by PINKI DESIGN
Interior Decoration and Implementation TATS Artists Interior Decoration and Implementation

主题名称 大艺术家·春光赋
项目名称 广州天河金茂广场会所销售中心
客户名称 中国金茂
户　　型 会所
设计面积 950平方米
项目地点 中国广州
创意出品 PINKI DESIGN品伊设计机构
软装设计与实现 TATS大艺术家软装设计与实现

GREAT ARTIST - SPRING RHAPSODY

大艺术家·春光赋

中国金茂集团 中国 广州

Sensibility is gradual, sensuality is expansion. With a one-time intersection of design, art and literature, waiting for you while standing in the light, witnessing and enjoying the artistic experience of the cycle and changes of the four seasons.
— Danfu Lau

理性是渐进，感性是扩展
用一次设计、艺术、文学的交汇
站在有光的地方等你
见证享受四季变化轮回中的
美学体验之旅
——刘卫军

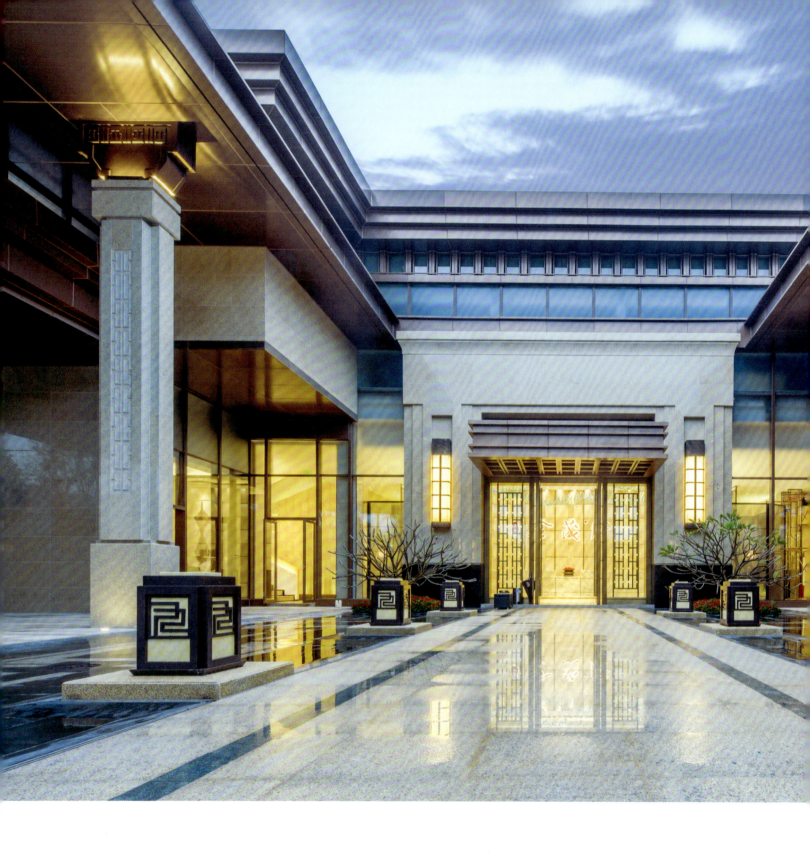

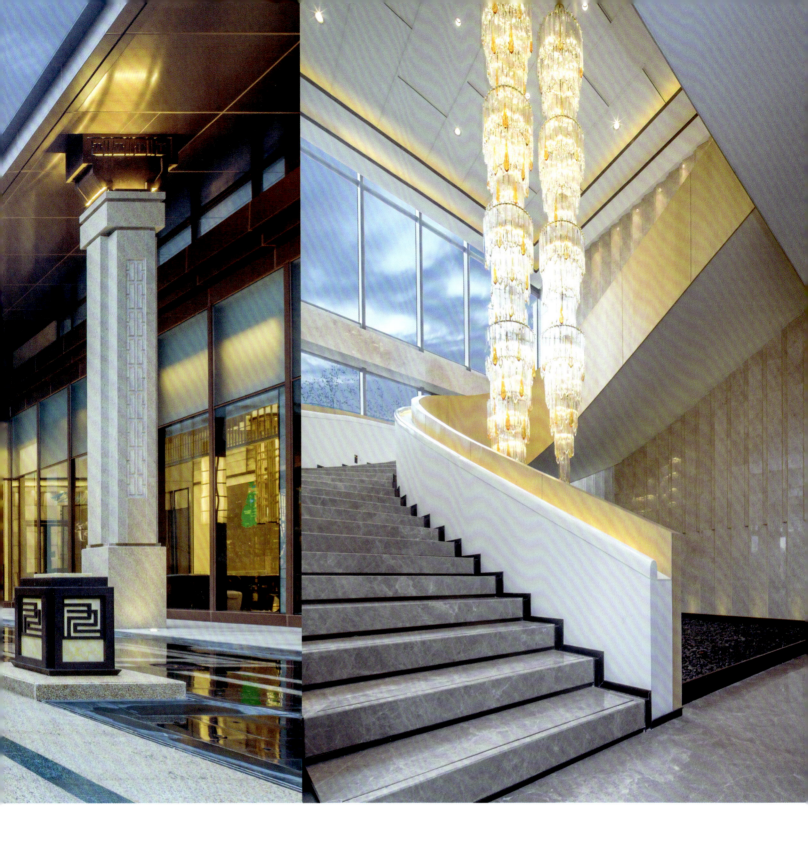

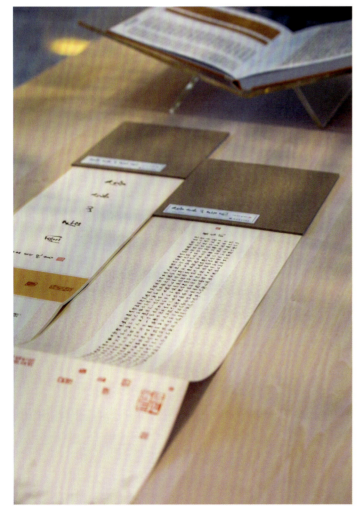

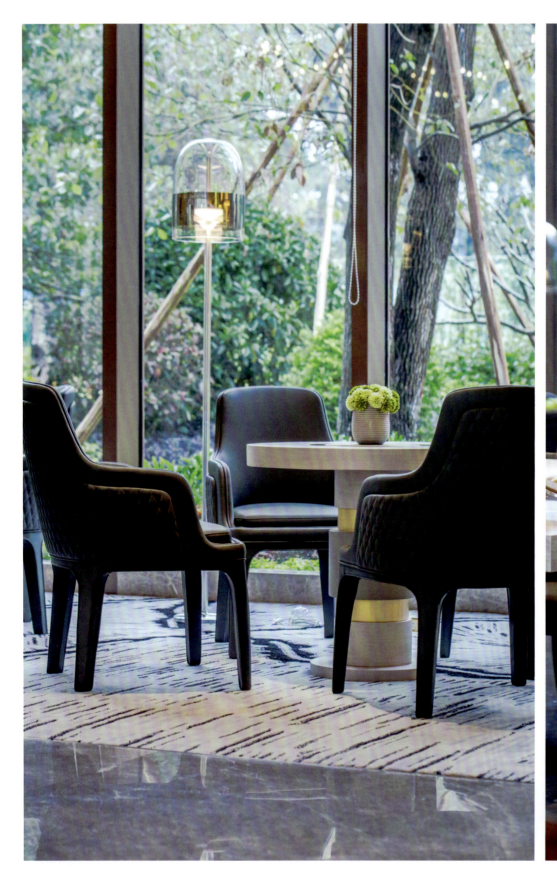

GREAT AR

ST - PERCH

Topic	Great Artist - Perch
Project	Taishan Dongfang Huacheng Court Clubhouse
Customer	Taishan Juyuan Real Estate Development Co., Ltd.
Unit	Clubhouse
Size	1070 square meters
Location	Taishan
Created by	PINKI DESIGN
Interior Decoration and Implementation	TATS Artists Interior Decoration and Implementation
Primary Materials	Marble, wooden finishes, tapestry, metal

主题名称	大艺术家 · 栖
项目名称	台山市东方华城苑会所
客户名称	台山市聚源房地产开发有限公司
户　　型	会所
设计面积	1070平方米
项目地点	中国台山
创意出品	PINKI DESIGN品伊设计机构
软装设计与实现	TATS大艺术家软装设计与实现
主要材料	大理石、木饰面、墙布、金属

ARTIST THINKING

GREAT ARTIST - PERCH

大艺术家·栖

台山市聚源房地产开发有限公司　中国 台山

Decipher the living arrangements of Oriental courtyard residence and the traditional humanistic philosophy of life, construct a beautiful scenery to interpret the "perch" approach as well as the selection of attitude.
— Danfu Lau

解读东方庭院住居的生活状态
与传统人文的生活哲学
构建一处美好的场景
来诠释"栖"的方式与选择的态度
——刘卫军

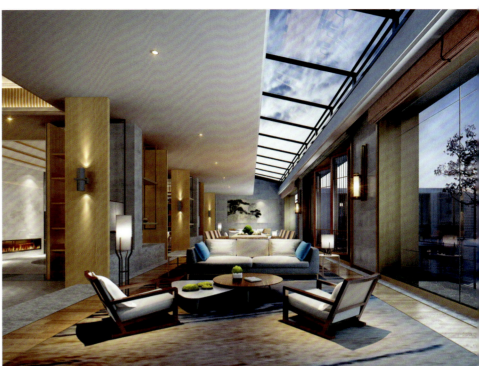

PINKI DESIGN

TIME

TSIDE

PINKI DESIGN

Topic Time Outside
Project Guangzhou Nansha Jinmao Bay Villa Model Suite
Customer China Jinmao
Unit Unit D12
Size 585 square meters
Location Guangzhou
Created by PINKI DESIGN
Interior Decoration and Implementation TATS Artists Interior Decoration and Implementation
Primary Materials Marble, wooden finishes, tapestry, art glass, metal
Project Awards 2016 Taiwan TAKAO Interior Design Awards Commercial Space Category Award, APDC Praise/Creativity 15/16 Asia Pacific Interior Design Elite Invitational Award.

主题名称 遇外时光
项目名称 广州南沙金茂湾商墅样板房
客户名称 中国金茂
户　　型 D12户型
设计面积 585平方米
项目地点 中国广州
创意出品 PINKI DESIGN品伊设计机构
软装设计与实现 TATS大艺术家软装设计与实现
主要材料 大理石、木饰面、墙布、艺术玻璃、金属
项目获奖 2016台湾TAKAO室内设计大赏商业空间类优选奖、APDC赞颂·创造力2015/2016亚太室内设计精英邀请赛佳作奖。

Time Outside

遇外时光

中国金茂集团　中国 广州

All emotions and settings are inseparable from time, but one day when the time is allowed to be stopped, all that has happened are then a story that has happened at another space-time that cannot be remembered.
— Danfu Lau

所有情与境都离不开时间的关系
但有一天 时间静止了
所有的一切
就是在另一个时空发生了一场无法
记忆的故事
——刘卫军

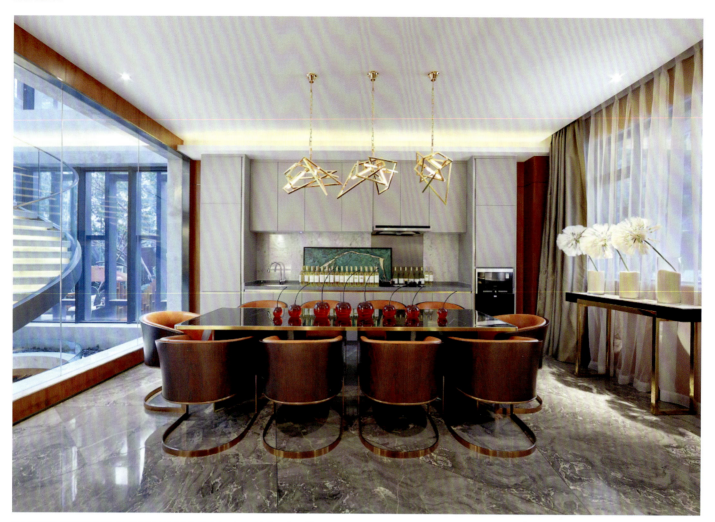

PINKI DESIGN

THE CON
ESTHE

MPORARY

ICIAN

PINKI DESIGN

Topic	The Contemporary Esthetician	主题名称	大时代美学家
Project	Foshan Lake Ludao Project Villa Model Suite	项目名称	佛山绿岛湖项目别墅样板间
Customer	China Jinmao	客户名称	中国金茂
Unit	Unit 330	户　　型	330户型
Size	850 square meters	设计面积	850平方米
Location	Foshan	项目地点	中国佛山
Created by	PINKI DESIGN	创意出品	PINKI DESIGN品伊设计机构
Interior Decoration	TATS Artists Interior Decoration and Implementation	软装设计	TATS大艺术家软装设计与实现
Primary Materials	Marble, wooden finishes, tapestry, metal, art glass, wood flooring	主要材料	大理石、木饰面、墙布、金属、艺术玻璃、木地板

THE CONTEMPORARY ESTHETICIAN

大时代美学家

中国金茂集团　中国 佛山

Although the times have created all the good things in the world, yet we must still pay tribute to the many times that laughter has echoed throughout the space, the wonderful and touching family moments, as well as the warmth and contentment of companionship.
— Danfu Lau

尽管时代造就了人世间所有的美好
但依然要致敬
那曾经多少次欢声笑语
回响过的空间中
我们家庭时光那般的美妙与感动
还有相依陪伴的那种温暖与满足
——刘卫军

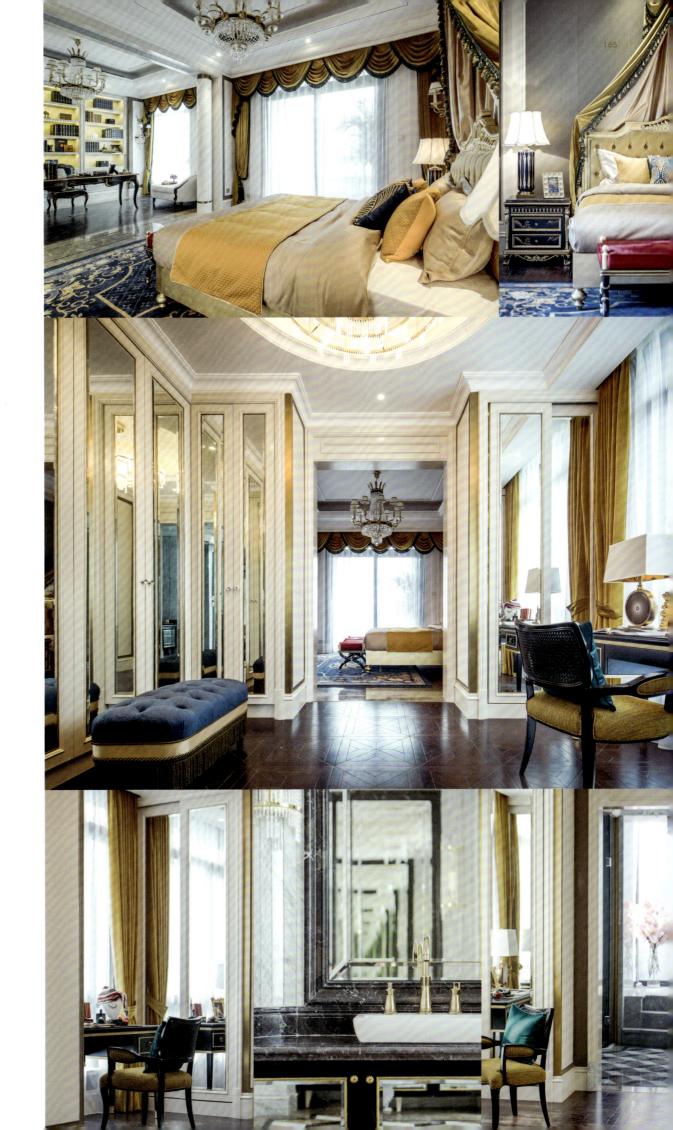

BEAUTIFUL

WINTER   F

WORLD OF
 RY TALE

PINKI DESIGN

Topic   Beautiful World of Winter Fairy Tale
Project   Guanzhou Zhujiang Jinmao Mansion Clubhouse
Customer   China Jinmao
Unit   Clubhouse
Size   2669 square meters
Location   Guangzhou
Created by   PINKI DESIGN
Interior Decoration and Implementation   TATS Artists Interior Decoration and Implementation
Primary Materials   Marble, wooden finishes, tapestry, art glass, metal

主题名称   冬日童话里的美丽世界
项目名称   广州珠江金茂府会所
客户名称   中国金茂
户　　型   会所
设计面积   2669平方米
项目地点   中国广州
创意出品   PINKI DESIGN品伊设计机构
软装设计与实现   TATS大艺术家软装设计与实现
主要材料   大理石、木饰面、墙布、艺术玻璃、金属

BEAUTIFUL WORLD OF WINTER FAIRY TALE

冬日童话里的美丽世界

中国金茂集团　中国 广州

Infusing the spatial properties that has already transformed, creating borderless imagination of the setting, extending the fairy tales in the memories just like those of childhood.
— Danfu Lau

融合已变更的空间属性
创造无边界的情境想象
延续如童年追忆中的**童话故事**
——刘卫军

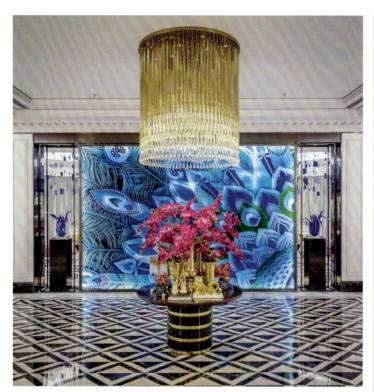

PINKI DESIGN

 KOI AND

BEGONIA

PINKI DESIGN

Topic	Koi and Begonia
Project	Guanzhou Zhujiang Jinmao Mansion
Customer	China Jinmao
Unit	Entrance lobby public space
Size	977 square meters
Location	Guangzhou
Created by	PINKI DESIGN
Interior Decoration and Implementation	TATS Artists Interior Decoration and Implementation
Primary Materials	Marble, wooden finishes, ceramic tiles, metal

主题名称	大鱼海棠
项目名称	广州珠江金茂府
客户名称	中国金茂
户　　型	入户大堂公共空间
设计面积	977平方米
项目地点	中国广州
创意出品	PINKI DESIGN品伊设计机构
软装设计与实现	TATS大艺术家软装设计与实现
主要材料	大理石、木饰面、瓷砖、金属

KOI AND BEGONIA

大鱼海棠

中国金茂集团　中国 广州

Beginning a journey of humanistic and natural aesthetics and realizing its combination with an exploration of special and artistic aesthetics. Return to a purely authentic and sincere view of life.
— Danfu Lau

开启一次人文自然美学的
旅行与空间艺术美学的探索
将二者结合实现
回归一种纯粹真实与真诚的生活观
——刘卫军

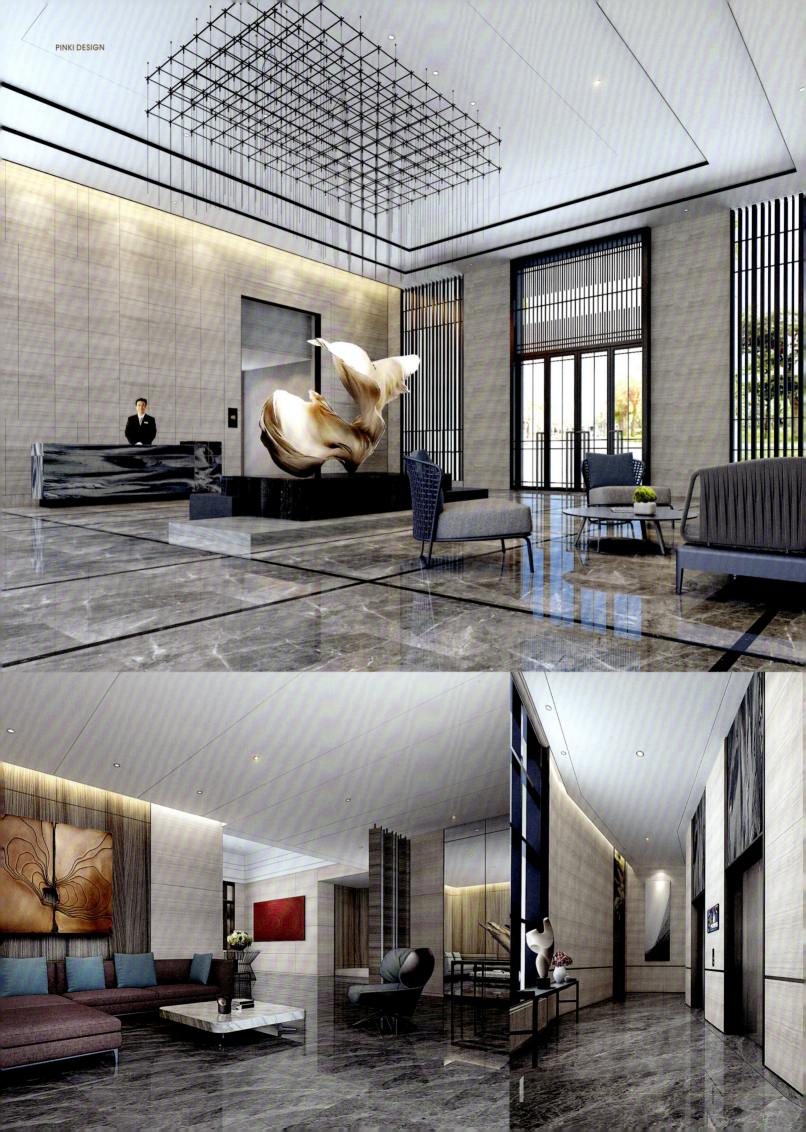

PINKI DESIGN

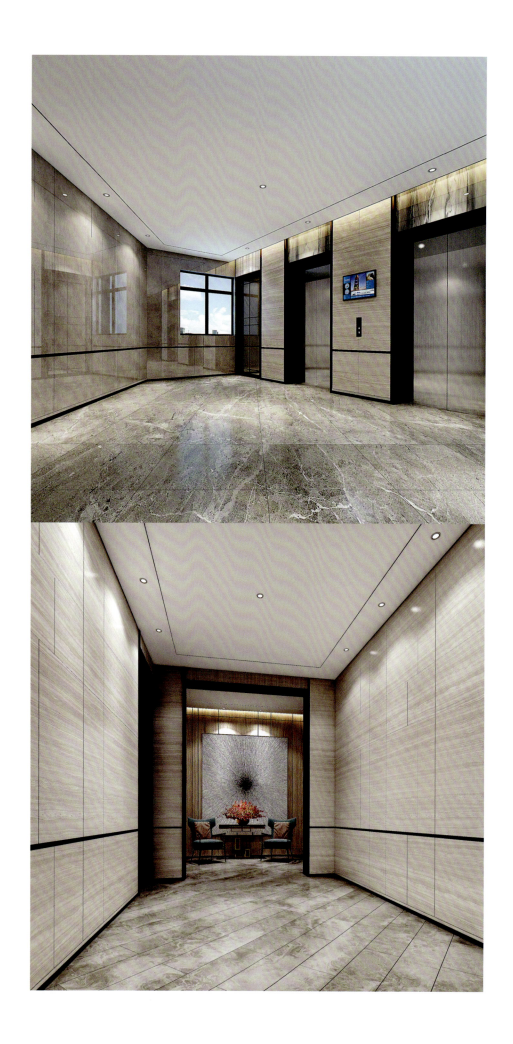

DIFFUS

TIME

PINKI DESIGN

Topic Diffused Time
Project Huaxia Tianjing Bay Villa Model Suite
Customer Yunnan Huaxia Real Estate Co., Ltd.
Unit Villa
Size 818 square meters
Location Kunming
Created by PINKI DESIGN
Interior Decoration TATS Artists Interior Decoration and Implementation
Primary Materials Marble, wooden finishes, tapestry, metal, art glass, wood flooring
Project Awards APDC Praise/Creativity 15/16 Asia Pacific Interior Design Elite Invitational Model Suite Room Space Silver Award, 4th China Space Design Summit Gold Award, 2015 China Interior Design Grand Prix (CIID Society Award) Bronze Award.

主题名称 唯漫时光
项目名称 华夏天璟湾别墅样板房
客户名称 云南华夏房地产有限公司
户　　型 别墅
设计面积 818平方米
项目地点 中国昆明
创意出品 PINKI DESIGN品伊设计机构
软装设计 TATS大艺术家软装设计与实现
主要材料 大理石、木饰面、墙布、金属、艺术玻璃、木地板
项目获奖 APDC赞颂·创造力2015/2016亚太室内设计精英邀请赛样板房空间类银奖、2016第四届中国营造空间设计峰会金奖、2015中国室内设计大奖赛（CIID学会奖）铜奖。

DIFFUSED TIME

唯漫时光

云南华夏房地产有限公司　中国 昆明

Champion nature, let your soul return to its source, a wonderful enjoyment of letting your heart flow, an exuberant experience of basking in humanity; time flies, let us not waste the diffused time and nature. Guide the details of spatial design with the spirit of ingenuity, construct a space for experiencing life rich in quality and the contemporary.
— Danfu Lau

崇尚自然，让心灵回归本源
这是一种心随自然的美好享受
以及一场身浸人文的华丽体验
白驹过隙，唯漫时光与自然不可辜负！
用匠心传承的精神
去传导空间的设计细节
营造一个富有品质与时代感的
生活体验空间！
——刘卫军

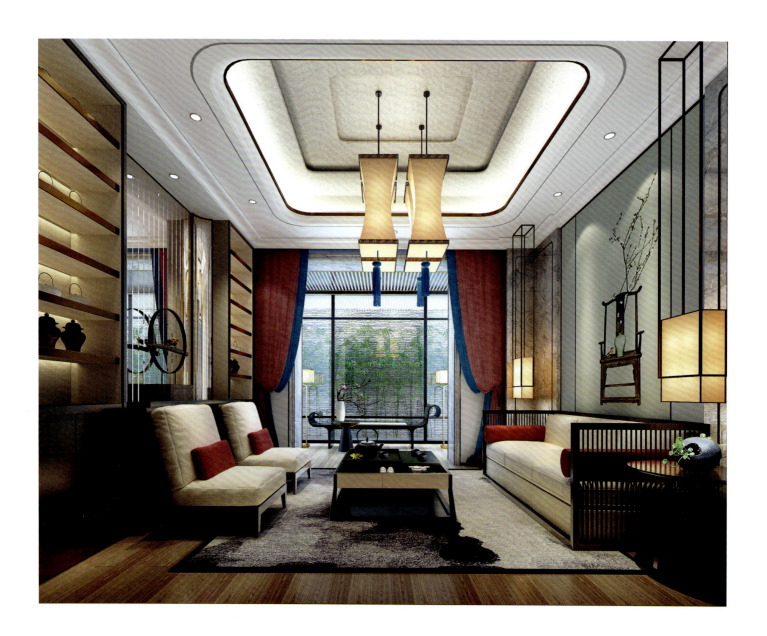

   A Z

R E

PINKI DESIGN

Topic	Azure
Project	Litang Shoushan Hot Spring Park Hotel
Customer	Hulu Island Litang Tourism Real Estate Development Co., Ltd.
Unit	Resort
Size	36,000 square meters
Location	Huludao
Created by	PINKI DESIGN
Interior Design	PINKI DESIGN US IARI LiuWeiJun Design Firm
Primary Materials	Marble, wooden finishes, tapestry, ceramic tiles, metal
Project Awards	2016 Innodesign Prize Space Design Silver Award, APDC Praise/Creativity 15/16 Asia Pacific Interior Design Elite Invitational Hotel Space Category Bronze Award, 4th China Space Design Summit Gold Award, 2015 Macau Design Exhibition "Golden Lotus" Cup International Design Master Invitational Silver Award, IHFO 2013 Chinese and Foreign Hotel Forum Platinum Award, 2013 China Interior Design Grand Prix (CIID Society Award) Annual Program Bronze Award.

主题名称	青蓝幽境
项目名称	丽汤·首山温泉公园酒店
客户名称	葫芦岛丽汤旅游房地产开发有限公司
户　　型	旅游度假区
设计面积	3.6万平方米
项目地点	中国葫芦岛
创意出品	PINKI DESIGN品伊设计机构
室内建筑规划	美国IARI刘卫军设计事务所
主要材料	大理石、木饰面、墙布、瓷砖、金属
项目获奖	INNODESIGN－2016法国双面神空间设计大奖银奖、APDC赞颂·创造力2015/2016亚太室内设计精英邀请赛酒店空间类铜奖、2016第四届中国营造空间设计峰会金奖、2015澳门设计联展"金莲花"杯国际设计大师邀请赛银奖、IHFO 2013中外酒店论坛白金奖、2013中国室内设计大奖赛（CIID学会奖）年度方案类铜奖。

PINKI DESIGN

CONN

SSEUR

PINKI DESIGN

Topic	Connoisseur	**主题名称**	鉴藏
Project	Dalian YiDa Miaoling Project	**项目名称**	大连亿达庙岭项目
Customer	YiDa Group	**客户名称**	亿达集团
Unit	Clubhouse	**户　　型**	会所
Size	3956 square meters	**设计面积**	3956平方米
Location	Dalian	**项目地点**	中国大连
Created by	PINKI DESIGN	**创意出品**	PINKI DESIGN品伊设计机构
Interior Decoration	TATS Artists Interior Decoration and Implementation	**软装设计**	TATS大艺术家软装设计与实现
Primary Materials	Marble, wooden finishes, tapestry, coating, art glass, wood flooring	**主要材料**	大理石、木饰面、墙布、涂料、艺术玻璃、木地板

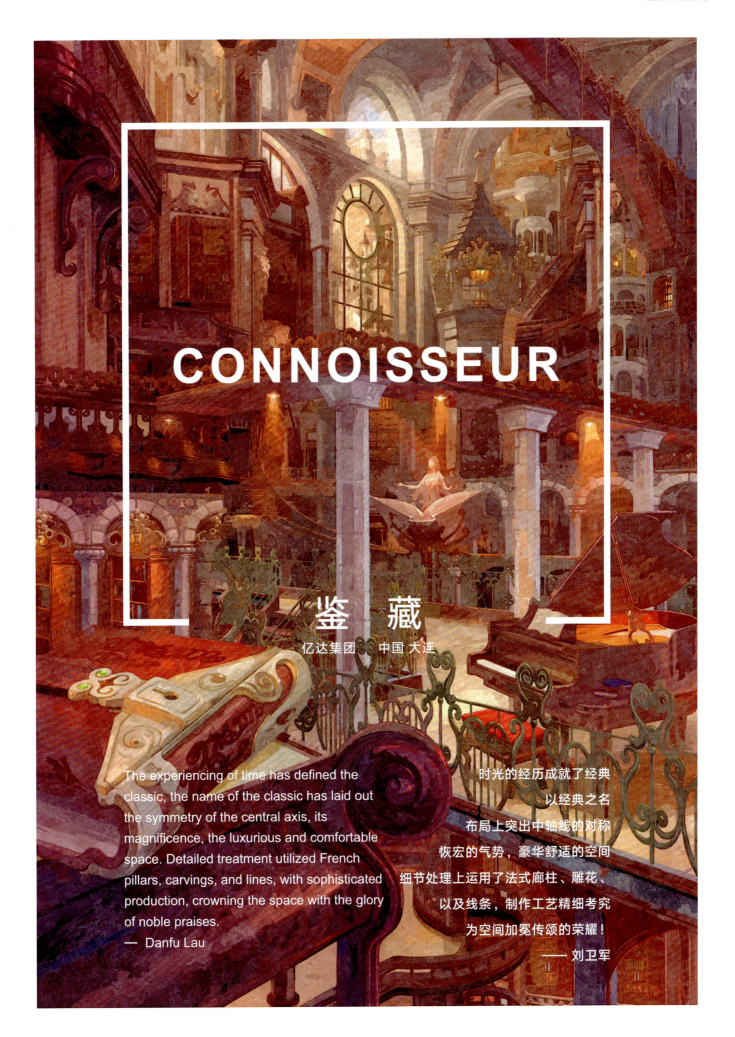

CONNOISSEUR

鉴 藏

亿达集团　中国 大连

The experiencing of time has defined the classic, the name of the classic has laid out the symmetry of the central axis, its magnificence, the luxurious and comfortable space. Detailed treatment utilized French pillars, carvings, and lines, with sophisticated production, crowning the space with the glory of noble praises.
— Danfu Lau

时光的经历成就了经典
以经典之名
布局上突出中轴线的对称
恢宏的气势，豪华舒适的空间
细节处理上运用了法式廊柱、雕花、
以及线条，制作工艺精细考究
为空间加冕传颂的荣耀！

——刘卫军

UNCH

RTED

PINKI DESIGN

Topic	Uncharted	主题名称	圆中秘境
Project	Pinghu Wanfu Park Clubhouse	项目名称	平湖万孚尊园会所
Customer	Pinghu Yayuan Property Co., Ltd.	客户名称	平湖雅园置业有限公司
Unit	Clubhouse	户　　型	会所
Size	2795 square meters	设计面积	2795平方米
Location	Hangzhou	项目地点	中国杭州
Created by	PINKI DESIGN	创意出品	PINKI DESIGN品伊设计机构
Interior Decoration	TATS Artists Interior Decoration and Implementation	软装设计	TATS大艺术家软装设计与实现
Primary Materials	Marble, wooden finishes, tapestry, art coating	主要材料	大理石、木饰面、墙布、艺术涂料

CORRIDOR

OF   TIME

PINKI DESIGN

Topic	Corridor of Time	主题名称	时光走廊
Project	Duobianhui Private Custom Branded Club	项目名称	多边荟私人定制品牌俱乐部
Customer	Beijing Duobianhui Investment Co., Ltd.	客户名称	北京多边荟投资有限公司
Unit	Clubhouse	户　　型	会所
Size	2,809 square meters	设计面积	2809平方米
Location	Beijing	项目地点	中国北京
Created by	PINKI DESIGN	创意出品	PINKI DESIGN品伊设计机构
Interior Decoration	TATS Artists Interior Decoration and Implementation	软装设计	TATS大艺术家软装设计与实现
Primary Materials	Marble, wooden finishes, carpet, art glass, wallpaper	主要材料	大理石、木饰面、地毯、艺术玻璃、墙纸

CORRIDOR OF TIME

时光走廊

北京多边荟投资有限公司　　中国 北京

The core value of design is not only to pursue the appeal of forms in space-time; if we are able to resolve the functions of spatial use for the objective of design, I believe the design value reflected shall be even more convincing.
— Danfu Lau

设计的核心价值
不仅是追求时间与空间形式的感染力
如果我们能够以空间使用的
功能为设计目的
相信设计的价值体现更具说服力
——刘卫军

ORIENTA

CHARM

PINKI DESIGN

Topic	Oriental Charm	主题名称	韵魅东方
Project	Shiou Shangjiangcheng Clubhouse	项目名称	世欧·上江城会所
Customer	Fuzhou Zhenro Group	客户名称	福州正荣集团
Unit	Clubhouse	户　　型	会所
Size	3470 square meters	设计面积	3470平方米
Location	Fuzhou	项目地点	中国福州
Created by	PINKI DESIGN	创意出品	PINKI DESIGN品伊设计机构
Interior Decoration and Implementation	TATS Artists Interior Decoration and Implementation	软装设计与实现	TATS大艺术家软装设计与实现
Primary Materials	Marble, wooden finishes, antique bricks, mosaic, wood flooring	主要材料	大理石、木饰面、仿古砖、马赛克、木地板
Project Awards	2015 United Nations 70+ Chinese Contemporary Art and Creative Design Achievement Award, Jintang Best Club Space Award.	项目获奖	2015年联合国70+华人当代艺术·创意设计成就荣誉奖、金堂最佳会所空间奖。

ORIENTAL CHARM

韵魅东方

福州正荣集团　中国 福州

This is an impressionistic design of Oriental humanistic aesthetics, allowing the innate memories of a Easterner, such as colors, shapes, and forms, to be laid out in space, realizing a resonance between men and the environment, looking for a sense of belonging that resembles zero distance.
— Danfu Lau

这是一次东方人文美学印象设计
让东方人与生俱来的记忆
如色彩、造型、形式围合等铺陈在空
间中实现人与环境关系的共鸣
寻求达成一种相似相随的零距离的
归属感
——刘卫军

 HAO SHE

G   YIN   YI

PINKI DESIGN

Topic	Hao Sheng Yin Yi
Project	Changsha Zhongjian Hotel Lake Meixi No. 1 Villa Model Suite
Customer	CSCEC
Unit	Unit 230
Size	340 square meters
Location	Changsha
Created by	PINKI DESIGN
Interior Decoration	TATS Artists Interior Decoration and Implementation
Primary Materials	Marble, wooden finishes, tapestry, leather, art glass, wood flooring
Project Awards	2013 Jintang Award: Annual Top Ten Model Suite Award

主题名称	豪笙印溢
项目名称	长沙中建梅溪湖壹号别墅样板间
客户名称	中国中建地产
户　　型	230户型
设计面积	340平方米
项目地点	中国长沙
创意出品	PINKI DESIGN品伊设计机构
软装设计	TATS大艺术家软装设计与实现
主要材料	大理石、木饰面、墙布、皮革、艺术玻璃、木地板
项目获奖	2013金堂奖年度十佳样板间大奖

ARTIST THINKING

HAO SHENG YIN YI

豪笙印溢

中国中建地产　　中国 长沙

Attempt to gently knock on the gate of the inherent design concepts and models of the East, allow the breeze of fashion and trend to flow past you, this could be a new Oriental exploration.
—— Danfu Lau

尝试轻轻敲开
东方固有设计理念模式的门扇
让时尚流行的清风
吹拂一下
也可能是一种新东方的探索
—— 刘卫军

DRUNKEN

THE   P

ARDEN OF
ACHES

PINKI DESIGN

Topic	Drunken Garden of the Peaches	主题名称	醉慕桃源
Project	Block 306, Nanhu Lot, Lotus Pond Moonlight Park Clubhouse Project	项目名称	南湖地段306地块荷塘月色公园会所项目
Customer	Wenzhou Zhonglianghe Real Estate Co., Ltd.	客户名称	温州市中梁和置业有限公司
Unit	Clubhouse	户型	会所
Size	1022 square meters	设计面积	1022平方米
Location	Wenzhou	项目地点	中国温州
Created by	PINKI DESIGN	创意出品	PINKI DESIGN品伊设计机构
Interior Decoration	TATS Artists Interior Decoration and Implementation	软装设计	TATS大艺术家软装设计与实现
Primary Materials	Marble, wooden finishes, tapestry, carpet	主要材料	大理石、木饰面、墙布、地毯

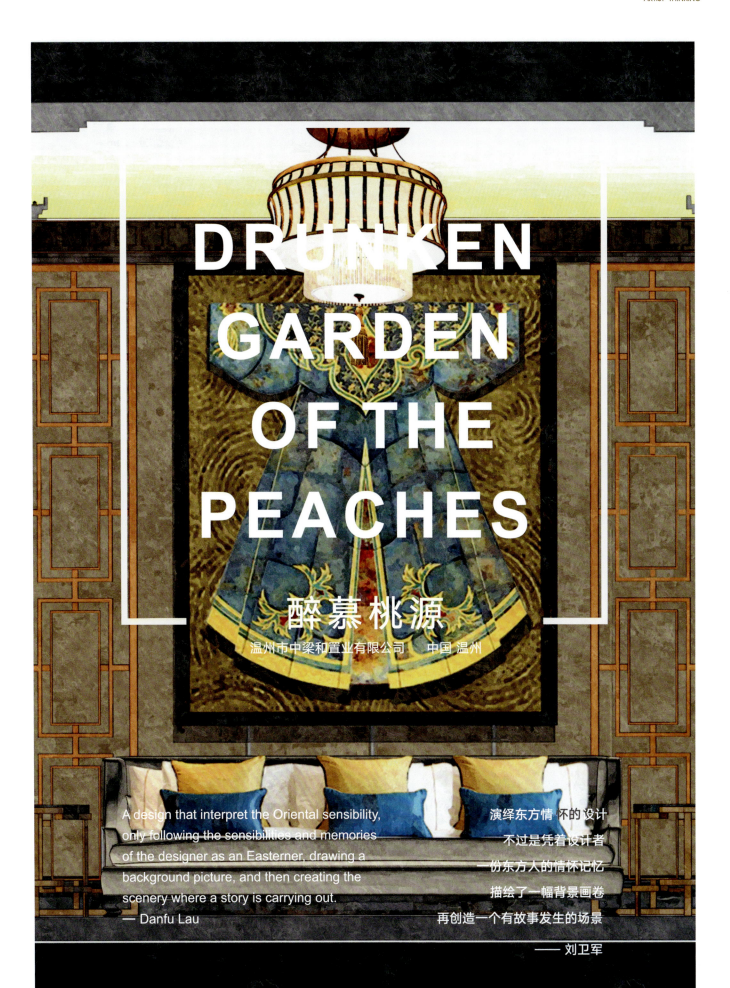

DRUNKEN GARDEN OF THE PEACHES

醉慕桃源

温州市中梁和置业有限公司　中国 温州

A design that interpret the Oriental sensibility, only following the sensibilities and memories of the designer as an Easterner, drawing a background picture, and then creating the scenery where a story is carrying out.
— Danfu Lau

演绎东方情怀的设计
不过是凭着设计者
一份东方人的情怀记忆
描绘了一幅背景画卷
再创造一个有故事发生的场景

—— 刘卫军

SENSE - TIM

TRAVELER

Topic	Sense - Time Traveler
Project	Xi'an Yangguang Jincheng Villa Clubhouse
Customer	Xi'an Guozhong Xingcheng Properties Limited
Unit	Clubhouse
Size	700 square meters
Location	Shenzhen
Created by	PINKI DESIGN
Interior Decoration and Implementation	TATS Artists Interior Decoration and Implementation
Primary Materials	Marble, wooden finishes, tapestry, cultured stone, wood flooring
Project Awards	2011 Jintang Award Annual Leisure Space Design Excellence Award, The 7th Golden Bund Award for Best Landscape Design.

主题名称	感·时光旅者
项目名称	西安·阳光金城别墅会所
客户名称	西安国中星城置业有限公司
户　　型	会所
设计面积	700平方米
项目地点	中国深圳
创意出品	PINKI DESIGN品伊设计机构
软装设计与实现	TATS大艺术家软装设计与实现
主要材料	大理石、木饰面、墙布、文化石、木地板
项目获奖	2011金堂奖年度休闲空间设计优秀奖、2012第七届金外滩奖、最佳景观设计大奖。

ARTIST THINKING

Sense - Time Traveler

感·时光旅者

西安国中星城置业有限公司　中国 深圳

The interior, architecture, garden flows with the heart and the setting, born from the backdrop. Create a painting of real-life landscape using a most primitive, simplistic design method without any deliberately modeled approach.
— Danfu Lau

室内、建筑、园林随心随势
随景而生
用一种最原始质朴
且没有刻意模式化的设计方式
创作一幅真实的生活风景画
——刘卫军

BAROQUE

OVE SONG

PINKI DESIGN

Topic	Baroque Love Song	主题名称	巴洛克恋曲
Project	Changsha CSCEC Lake Meixi No. 1 Villa Model Suite	项目名称	长沙中建梅溪湖壹号别墅样板间
Customer	CSCEC	客户名称	中国中建地产
Unit	Unit 280	户　　型	280户型
Size	507 square meters	设计面积	507平方米
Location	Changsha	项目地点	中国长沙
Created by	PINKI DESIGN	创意出品	PINKI DESIGN品伊设计机构
Interior Decoration and Implementation	TATS Artists Interior Decoration and Implementation	软装设计与实现	TATS大艺术家软装设计与实现
Primary Materials	Marble, wooden finishes, tapestry, leather, ceramic tiles, mosaic, wood flooring	主要材料	大理石、木饰面、墙布、皮革、瓷砖、马赛克、木地板
Project Awards	2013 CIDA China Interior Design Awards Annual Model Design Award	项目获奖	2013年CIDA中国室内设计大奖年度样板间设计奖

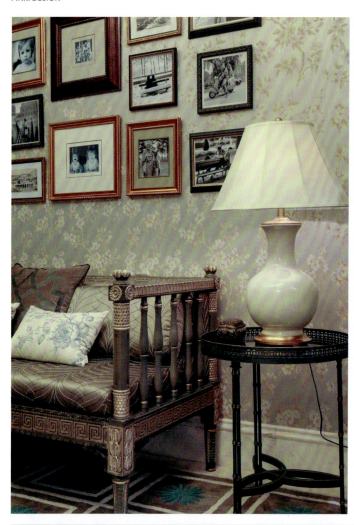

SONG OF

AND FU

FLOWERS

MOON

PINKI DESIGN

Topic	Song of Flowers and Full Moon	主题名称	花好月圆曲
Project	Yangguang City - Shanglin Fuyuan Villa Model Suite	项目名称	阳光城·上林赋苑别墅样板房
Customer	Shaanxi Shanglinyuan Investment & Development Co., Ltd.	客户名称	陕西上林苑投资开发有限公司
Unit	Terraced Villa 29 West Top Unit	户　　型	叠拼别墅29号西边上户
Size	250 square meters	设计面积	250平方米
Location	Xi'an	项目地点	中国西安
Created by	PINKI DESIGN	创意出品	PINKI DESIGN品伊设计机构
Interior Decoration	TATS Artists Interior Decoration and Implementation	软装设计	TATS大艺术家软装设计与实现
Primary Materials	Marble, wooden finishes, tapestry, wood flooring	主要材料	大理石、木饰面、墙布、木地板
Project Awards	The Annual Top 10 Model Room Of JinTangPrize	项目获奖	2012金堂奖年度十佳样板间大奖

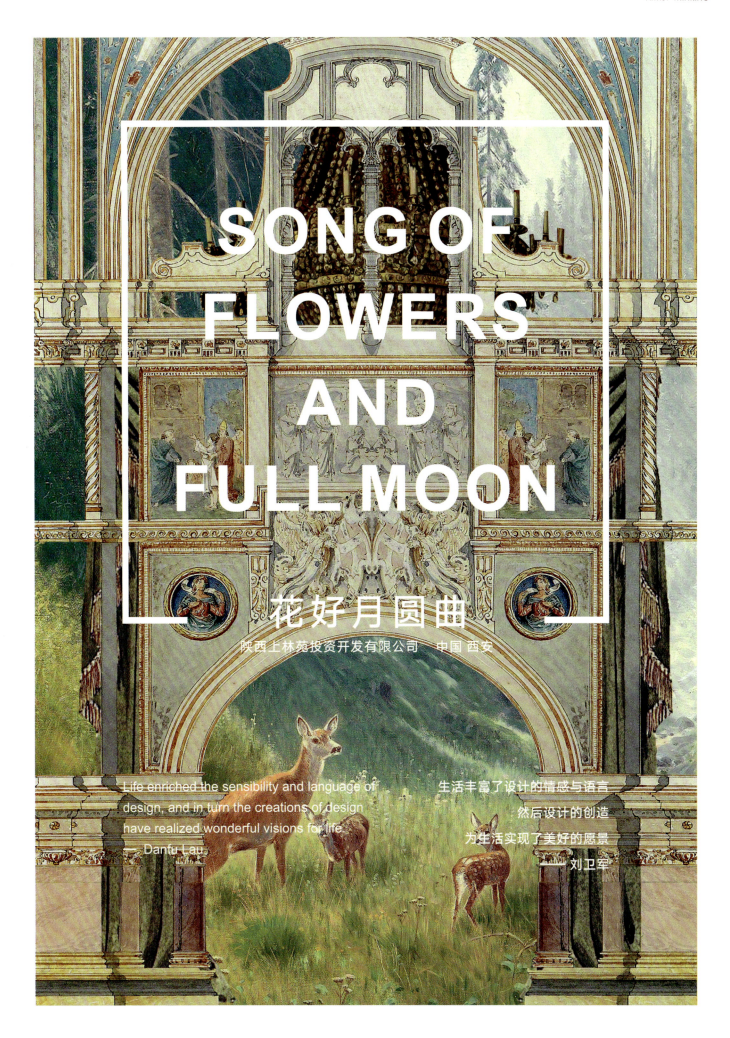

WAVES O

FLOWER

PINKI DESIGN

Topic	Waves of Flower
Project	Xinjiang China Airlines Emerald City Third Phase Villa Model Suite
Customer	China Airlines Real Estate
Unit	6202 Building No. 3 unit C
Size	304 square meters
Location	Xinjiang
Created by	PINKI DESIGN
Interior Decoration	TATS Artists Interior Decoration and Implementation
Primary Materials	Marble, wooden finishes, tapestry, antique bricks, mosaic, wood flooring
Project Awards	2013 CIDA China Interior Design Award Annual Model Suite Design Award.

主题名称	花漾漫波
项目名称	新疆中航翡翠城三期别墅样板间
客户名称	中国·中航地产
户　　型	6202栋3号C户型
设计面积	304平方米
项目地点	中国新疆
创意出品	PINKI DESIGN品伊设计机构
软装设计	TATS大艺术家软装设计与实现
主要材料	大理石、木饰面、墙布、仿古砖、马赛克、木地板
项目获奖	2013CIDA中国室内设计大奖年度样板间设计提名奖

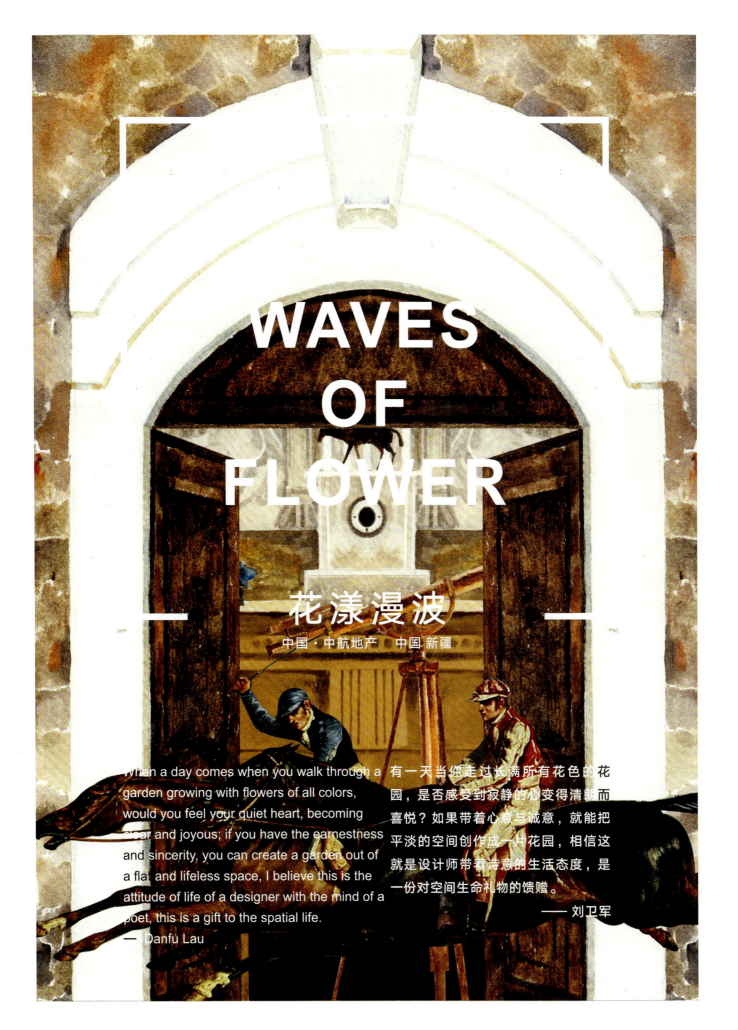

WAVES OF FLOWER

花漾漫波

中国·中航地产 中国 新疆

When a day comes when you walk through a garden growing with flowers of all colors, would you feel your quiet heart, becoming clear and joyous; if you have the earnestness and sincerity, you can create a garden out of a flat and lifeless space, I believe this is the attitude of life of a designer with the mind of a poet, this is a gift to the spatial life.
—— Danfu Lau

有一天当您走过长满所有花色的花园，是否感受到寂静的心变得清朗而喜悦？如果带着心意与诚意，就能把平淡的空间创作成一片花园，相信这就是设计师带着诗意的生活态度，是一份对空间生命礼物的馈赠。
——刘卫军

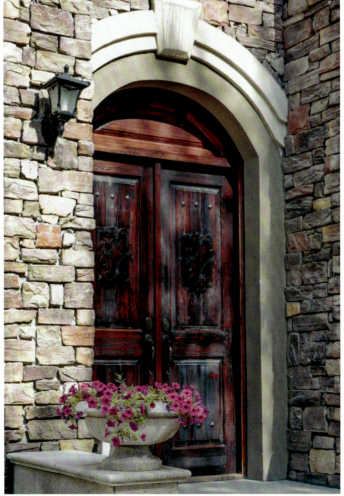

WATER

GARDEN

Topic   Water Garden
Project   Xinjiang China Airlines Emerald City Villa Model Suite
Customer   China Airlines Real Estate
Unit   6202 Building Villa Unit B
Size   290 square meters
Location   Xinjiang
Created by   PINKI DESIGN
Interior Decoration   TATS Artists Interior Decoration and Implementation
Primary Materials   Marble, wooden finishes, tapestry, art coating, wood flooring, antique bricks

主题名称   水榭听香
项目名称   新疆中航翡翠城别墅样板房
客户名称   中国·中航地产
户　　型   6202栋别墅B户型
设计面积   290平方米
项目地点   中国新疆
创意出品   PINKI DESIGN品伊设计机构
软装设计   TATS大艺术家软装设计与实现
主要材料   大理石、木饰面、墙布、艺术涂料、木地板、仿古砖

ARTIST THINKING

WATER GARDEN

水榭听香
中国·中航地产　中国 新疆

As long as there are stories in space, all those beautiful sceneries, you can hear its fragrance even when your eyes are closed
—— Danfu Lau

只要空间里有故事
所有美好的情境
当你闭上双眼也能感受到它的芳香
—— 刘卫军

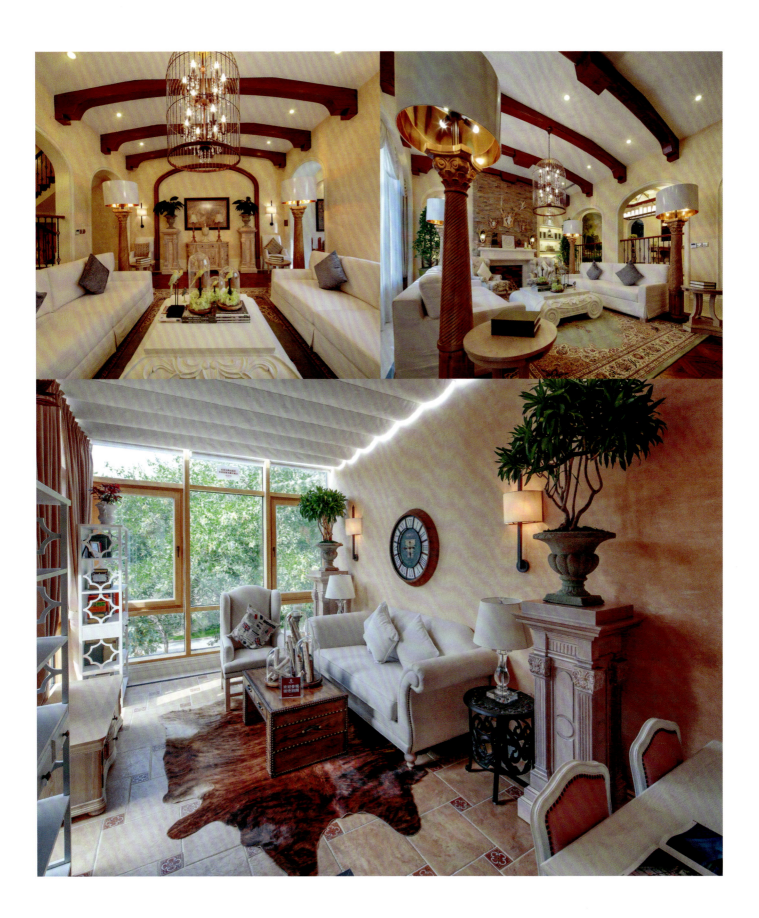

 SPLENDOR

F   ANDORRA

PINKI DESIGN

Topic	Splendor of Andorra	主题名称	安道拉的璀璨
Project	Yangguang City - Shanglin Fuyuan Villa Model Suite	项目名称	阳光城·上林赋苑别墅样板房
Customer	Shaanxi Shanglinyuan Investment & Development Co., Ltd.	客户名称	陕西上林苑投资开发有限公司
Unit	Township villa Building 62, first suite west	户　　型	联排别墅62号楼最西边第一边户
Size	300 square meters	设计面积	300平方米
Location	Xi'an	项目地点	中国西安
Created by	PINKI DESIGN	创意出品	PINKI DESIGN品伊设计机构
Interior Decoration	TATS Artists Interior Decoration and Implementation	软装设计	TATS大艺术家软装设计与实现
Primary Materials	Marble, wooden finishes, art coating, tapestry, antique bricks, wood flooring	主要材料	大理石、木饰面、艺术涂料、墙布、仿古砖、木地板

SPLENDOR OF ANDORRA

安道拉的璀璨

陕西上林苑投资开发有限公司　中国 西安

The transitioning and cycling of nature and its four seasons have painted the vast expanse of land with the colors of vibrant and splendid changes; the design bears the gift of nature, smelling the fragrance of the land, creating the space of life.
— Danfu Lau

大自然四季的交替轮回
为辽阔无垠的土地描绘出
斑斓变化的色彩
设计利用大自然的礼物
以及土地的芳香
创造出富有生命力的空间
——刘卫军

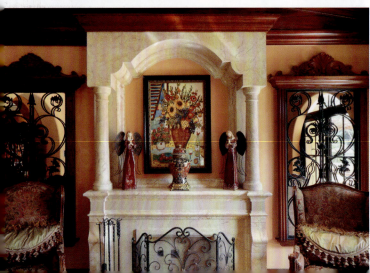

CALIFORNI

COTTAGE

PINKI DESIGN

Topic	California Cottage	主题名称	加州馨居
Project	Xinjiang China Airlines Emerald City Villa Model Suite	项目名称	新疆中航翡翠城别墅样板房
Customer	China Airlines Real Estate	客户名称	中国·中航地产
Unit	6101 Building Villa Unit C	户　　型	6101栋别墅C户型
Size	304 square meters	设计面积	304平方米
Location	Xinjiang	项目地点	中国新疆
Created by	PINKI DESIGN	创意出品	PINKI DESIGN品伊设计机构
Interior Decoration	TATS Artists Interior Decoration and Implementation	软装设计	TATS大艺术家软装设计与实现
Primary Materials	Marble, wooden finishes, tapestry, art coating, wood flooring	主要材料	大理石、木饰面、墙布、艺术涂料、木地板

 MORNING

OVE   SONG

PINKI DESIGN

Topic	Morning Love Song
Project	Yangguang City - Shanglin Fuyuan Villa Model Suite
Customer	Shaanxi Shanglinyuan Investment & Development Co., Ltd.
Uni	Terraced villa
Size	260 square meters
Location	Xi'an
Created by	PINKI DESIGN
Interior Decoration	TATS Artists Interior Decoration and Implementation
Primary Materials	Marble, wooden finishes, art coating, tapestry, wood flooring

主题名称	晨光恋曲
项目名称	阳光城·上林赋苑别墅样板房
客户名称	陕西上林苑投资开发有限公司
户　　型	叠拼别墅
设计面积	260平方米
项目地点	中国西安
创意出品	PINKI DESIGN品伊设计机构
软装设计	TATS大艺术家软装设计与实现
主要材料	大理石、木饰面、艺术涂料、墙布、木地板

MORNING LOVE SONG

晨光恋曲

陕西上林苑投资开发有限公司　中国 西安

Just to await that beam of light at dawn, because the most beautiful life is possible if there is light.
— Danfu Lau

只为等候晨间的那束光
因为有光就会有最美好的生活
——刘卫军

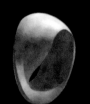

GREAT A
SPECTACU

CHITECT

AR   TIME

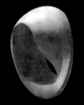

PINKI DESIGN

Topic	Great Architect, Spectacular Time	主题名称	大建筑家·曼妙时光
Project	China JinMao FoShan Green Island Lake Demonstration Area	项目名称	中国金茂佛山绿岛湖项目示范区
Customer	China Jinmao	客户名称	中国金茂集团
Size	450 square meters	设计面积	450平方米
Location	FoShan	项目地点	中国佛山
Created by	PINKI DESIGN	创意出品	PINKI DESIGN品伊设计机构
Primary Materials	Marble, wooden finishes, tapestry, art glass, metal	主要材料	大理石、木饰面、墙布、艺术玻璃、金属

GREAT ARCHITECT SPECTACULAR TIME

大建筑家·曼妙时光

中国金茂集团　中国 佛山

By dot, line, surface, build a combination of any form stereo space, design aesthetic effect of reason and sensibility, with the most contemporary language transfer. Reject, modular design approach, creating a pair of advancing with The Times, vivid landscapes plain, the flow of life.
— Danfu Lau

以点、线、面的方式，构建出任何形式组合而成的立体空间，传达理性与感性的设计美学效应，用最具时代感的语言传递。摒弃模式化的设计方式，创造一幅与时俱进的、生动平实的、流动的生活风景画。

———— 刘卫军

PINKI DESIGN

   ROYAL

NCENSE

PINKI DESIGN

Topic	Royal Incense	主题名称	御品倾香
Project	Dalian YiDa Azure Sky Model Suite	项目名称	大连亿达·青云天下样板房
Customer	YiDa Group	客户名称	亿达集团
Unit	Ua2 Bottom Unit	户　　型	Ua2底跃户型
Size	304 square meters	设计面积	304平方米
Location	Dalian	项目地点	中国大连
Created by	PINKI DESIGN	创意出品	PINKI DESIGN品伊设计机构
Interior Decoration	TATS Artists Interior Decoration and Implementation	软装设计	TATS大艺术家软装设计与实现
Primary Materials	Marble, wooden finishes, tapestry, wood flooring	主要材料	大理石、木饰面、墙布、木地板

ROYAL INCENSE

御品倾香

亿达集团　中国 大连

The details in design is the temperature expressed by space. The quality of the space is decided by the thinking of the designer regarding the integration between humanity and the environment.
— Danfu Lau

设计的细节
是空间表现的气质
空间品质取决于设计者对
人文环境融合统一的思考
——刘卫军

CONTEMPORA

YOUR NEI

y   NEW   HOME

HBORHOOD

PINKI DESIGN

Topic	Contemporary New Home Your Neighborhood
Project	New Perspective Humanistic Aesthetics Apartment
Customer	Shanghai
Unit	Apartment
Size	Multiple sets
Location	Various locations in China
Created by	PINKI DESIGN
Interior Decoration and Implementation	TATS Artists Interior Decoration and Implementation
Primary Materials	Marble, wooden finishes, wallpaper, wood flooring

主题名称	时代新居与你同邻
项目名称	新视角人文美学公寓
客户名称	中国中海集团
户　　型	公寓
设计面积	多套
项目地点	中国各地
创意出品	PINKI DESIGN品伊设计机构
软装设计与实现	TATS大艺术家软装设计与实现
主要材料	大理石、木饰面、墙纸、木地板

ARTIST THINKING

CONTEMPORARY NEW HOME YOUR NEIGHBORHOOD

时代新居与你同邻

中国中海集团　中国 各地

When the times are changing with a rapid rhythm, all of our sensory DNA also changes with the times; but there is a warm sensuality and a temperature of passion during an affectionate embrace that remains the same. You might be able to learn to cherish this moment and reminisce the pure feelings and memories of childhood.
— Danfu Lau

当时代日新月异地发展变化时，所有的感官基因，也会因时代的变化而改变，但有份温暖的感知，以及深情拥抱时热恋的温度，却依旧不变。这个时候你可能更懂得珍惜，重温孩童般最纯粹的情怀记忆。

海棠花未眠
HAI TANG HUA WEI MIAN

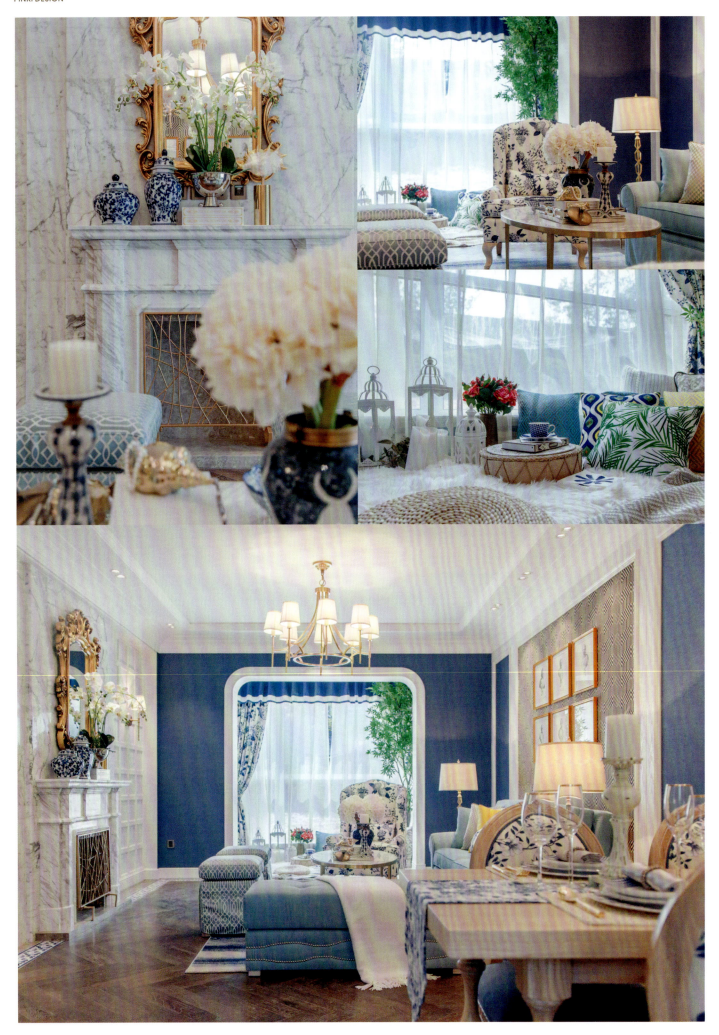

光之浴
GUANG ZHI YU

漫咖时光
MAN KA SHI GUANG

菲拉格慕
FEI LA GE MU

方糖
物语 FANG TANG
WU YU

曲悦风尚居
QU YUE FENG SHANG JU

LEGEND
AND

OF LOVE

REAM

PINKI DESIGN

Topic	Legend of Love and Dream	主题名称	爱与梦传奇的地方
Project	Lake Songshan Junyi College	项目名称	中国松山湖俊逸书院
Customer	PINKI International Creative Group	客户名称	PINKI品伊国际创意集团
Unit	Overall architecture and landscaping	户　　型	整体建筑与园林规划
Size	13,340 square meters	设计面积	20亩
Location	Lake Songshan	项目地点	中国松山湖
Created by	PINKI DESIGNPINKI DESIGN US IARI LiuWeiJun Design Firm	创意出品	PINKI DESIGN美国IARI刘卫军设计事务所
Interior Decoration and Implementation	TATS Artists Interior Decoration and Implementation	软装设计与实现	TATS大艺术家软装设计与实现
Primary Materials	Marble, wooden finishes, art coating, ceramic tiles, red bricks, bark, straw, reed	主要材料	大理石、木饰面、艺术涂料、瓷砖、红砖、树皮、稻草、芦苇杆

ARTIST THINKING

LEGEND OF LOVE AND DREAM

爱与梦传奇的地方

PINKI品伊国际创意集团　中国 松山湖

Everything are mere memories, yet all memories are for the sake of a sweet and tender smile covered in sweat from running, the design is also a place that allows the soil to be burning with the weeds.
— Danfu Lau

所有的一切只是一场记忆
所有的回忆却只是为了奔跑时
满脸汗滴的一张稚嫩笑脸
设计也就是为了一个让泥土伴着
荒草焚烧的地方
———— 刘卫军

PINKI DESIGN

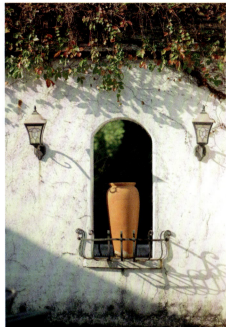

奖项列表

1998年　荣获深圳建设投资控股集团"优秀设计师"及"青年岗位能手青年"称号，深圳市装饰设计大赛第一名。

1999年　深圳科技大厦高新科技成果展览馆设计，荣获深圳市样板工程奖。

2000年　参与中华世纪坛室内设计，荣获中国最高设计工程鲁班奖。

2001年　参与中国建筑学会室内设计分会举办的中国室内设计大赛，荣获优秀奖及四项入围奖。

2002年　参与中国建筑学会室内设计分会举办的中国室内设计大赛，荣获优秀奖、三等奖及七项入围奖；同年荣获华南地区室内设计大赛第一名及最佳创意奖；也是这一年他获得了人生中第一个最重要的奖项——APIDA亚太室内设计大赛冠军奖。

2003年　第三次参与中国建筑学会室内设计分会举办的中国室内设计大赛，获得三等奖及八项入围奖；同年荣获APIDA亚太室内设计大赛荣誉大奖（中国区会所类唯一获奖者）。

2004年　荣获大陆、台湾、香港两岸三地室内设计大赛工程类二等奖、深圳建筑装饰集团设计研究院特别贡献奖；第四次参与中国建筑学会室内设计分会举办的中国室内设计大赛，获得三项大奖，并荣获全国商用空间设计大奖赛一等奖。

2005年　荣获IAID最具影响力中青年设计师、深圳市首届室内设计文化节工程及住宅类两项大奖、BDCI国际商年展网络最具人气奖、中国室内设计20年推动大奖、CIID学会贡献奖。

2006年　荣获"IDCFC城市荣誉杰出室内设计师"称号、中国十佳住宅设计师，第六次参与中国建筑学会室内设计分会举办的中国室内设计大赛并获得优秀奖，荣获中国（上海）国际建筑及室内设计节最佳居住空间奖、最佳陈设奖、最佳办公空间提名奖，于北京人民大会堂被授予"2004—2006年中国百名优秀室内建筑师"荣誉称号。

2007年　荣获第二届（博鳌）中国居家论坛"中国室内设计师风云人物"称号、中国（上海）国际建筑及室内设计节最佳概念设计提名奖，论文《怎样平衡环保意识》荣获中国室内设计年度优秀论文奖，荣获中国酒店设计大师赛优胜奖、第十五届APIDA亚太室内设计大赛会所类银奖（中国区会所类唯一荣获的最高奖项），被评为中国室内设计年度十大封面人物。

2008年　荣获中国（上海）国际建筑及室内设计节最佳饰品搭配奖提名奖、首届石材设计与艺术大赛创新设计奖，第七次参与中国建筑学会室内设计分会举办的中国室内设计大赛并获得优秀奖及佳作奖；同年他创立的PINKI品伊创意设计机构&美国IARI刘卫军设计师事务所被评为中国最佳住宅室内设计十强企业、中国最具价值室内设计十强企业。

2009年　荣获中国（上海）国际建筑及室内设计节最佳色彩运用优秀奖、现代装饰国际传媒奖年度样板空间大奖入围奖、搜狐焦点网"十大最具影响力牛博"、SOHU第二届设计师网络传媒、年度优秀博客奖，他带领团队获得现代装饰杯深圳设计师卡丁车邀请赛最具实力团队奖，被评为中国设计行业特高级研究员、全国设计行业首席专家荣获，全国首批设计行业优秀人才模范勋章、PINKI品伊创意设计机构&美国IARI刘卫军设计师事务所被评为中国国际设计艺术博览会"2009最具影响力设计机构",中国国际设计艺术博览会审定刘卫军总监为专家委员会委员,同时PINKI品伊创意设计机构&美国IARI刘卫军设计师事务所荣获中国国际设计艺术博览会银奖、中国时代新闻人物,获得中国建筑装饰协会举办的中国室内空间环境艺术设计大赛住宅空间优秀奖、会所空间三等奖、样板房二等奖、陈设空间二等奖,获得"新浪品鉴团专家"荣誉称号;代表作《藏青亭居》荣获雷士照明周刊杯中国照明设计大赛优秀奖,这一年中国建筑学会室内设计分会中国室内设计20年特授予刘卫军总监"学会贡献奖""优秀设计师"荣誉证书;PINKI品伊创意设计机构&美国IARI刘卫军设计师事务所荣获"深圳企业文化建设优秀单位"荣誉称号。

2010年 被列为和谐中国·2010年度影响力人物重点提名候选人;第五届中外酒店白金奖特授予其"中国酒店设计大师"荣誉称号,作品《永和坊》荣获金堂奖年度优秀作品奖,作品《爵士魅影》荣获现代装饰第八届国际传媒奖年度会所空间大奖入围奖,作品《梦遇香居》荣获现代装饰第八届国际传媒奖年度家居空间大奖入围奖,作品《梦遇香居》荣获国际空间设计大奖艾特奖最佳陈设艺术设计提名奖,PINKI品伊创意设计机构荣获"大中华区最具影响力设计机构"荣誉称号。

2011年 荣获第五届海峡两岸四地室内设计大赛最佳设计组合奖、杰出建筑空间金奖、"杰出设计师"称号;荣获新浪微博2010年度活跃奖;作品《云顶天阁》荣获中国(上海)国际建筑设计及室内设计节金外滩奖最佳景观设计入围奖;荣获"中国十大高端住宅设计师"荣誉称号;《美好家园》中国五周年暨美家之夜创美颁奖盛典在北京香港马会隆重举行,刘卫军先生被授予"室内设计大师"称号(其余分别为:邱德光、高文安、高意静、赖亚楠、杨明洁、紫香舸、安东、陈耀光、设计共和、明和文吉、蒋琼耳、石大宇、杨瑞丹、黄志达);荣获第六届中国国际设计艺术博览会2010—2011年度杰出设计师奖,作品《丛林里的诗意漫游》和《都市魅影》荣获CIID学会奖入选奖,第六届中外酒店白金奖特授予刘卫军先生"十大白金设计师"荣誉,刘卫军先生及团队荣获第六届海峡两岸四地室内设计大赛一金奖、两银奖、一铜奖,PINKI品伊创意集团十年专集《灵感塑造空间》荣获金堂奖年度最佳出版物提名奖,作品《感·时光旅者》荣获金堂奖年度休闲空间设计优秀奖,作品《和煦蕴逸》荣获金堂奖2011年度十佳样板间大奖。

2012年 作品《感·时光旅者》荣获2012第七届金外滩奖最佳景观设计大奖;个人荣获中国陈设艺术发展论坛年度室内陈设艺术先锋人物奖;作品《花好月圆曲》荣获金堂奖2012年度十佳样板间大奖。

2013年 作品《丽汤·温泉公园酒店》荣获第八届中外酒店白金奖2013年度最佳创意设计白金奖;作品《巴洛克恋曲》荣获2013年CIDA中国室内设计大奖2013年度样板间设计奖;作品《闲》荣获2013CIDA中国室内设计大奖2013年度样板间设计提名奖;作品《丽汤·温泉公园酒店》荣获中国室内设计学会奖2013年度方案类铜奖;作品《豪笙印溢》荣获金堂奖2013年度十佳样板间大奖;作品《侬本多情》荣获金堂奖2013年度样板间设计优秀奖;作品《怡心韵驿》荣获美居奖2013年度最美空间奖。拓展教育领域,刘卫军先生荣获中国室内装饰协会成立25周年「中国室内设计教育贡献奖。

2014年 《灵感塑造空间》系列丛书被国家图书馆收藏;第十七届中国室内设计大奖赛作品《雕刻时光》荣获住宅工程类入选奖;2014金堂奖《曲悦·风尚居》荣获年度优秀作品奖;2014第十七届中国室内设计大奖赛《曲悦·风尚居》荣获住宅工程类入选奖;荣获第十二届(2014)现代装饰国

际传媒奖年度杰出设计师大奖；2014第十届中国国际室内设计双年展《曲悦·风尚居》优秀奖；2014居然杯CIDA中国室内设计大奖《曲悦·风尚居》荣获居住空间·住宅设计奖；作品《曲悦·风尚居》荣获2014年美居奖北赛区"中国最美样板间"第一名；"世界向东"西湖软装艺术设计周授予刘卫军先生"2014世界向东传承助学导师和中国软装艺术金凤凰传承奖"。

2015年 《白日梦》荣获2015CWDA首届国际橱窗设计大赛颁奖典礼特别贡献奖；《唯漫时光》在2015CIID学会奖中获铜奖；《丽汤温泉公园酒店》在2015澳门"金莲花"杯国际设计大师邀请赛中获得银奖；荣获"2015中国家居时代精英设计师"荣誉；会所设计作品《韵魅东方》2015年入选联合国70+华人当代艺术·创意设计成就展；《再续雕刻时光》荣获2015地产设计美居奖中国最美样板间大奖；荣获CPC Award 2015 年度教育贡献人物奖；PINKI DESIGN荣获CPC Award2015 年度创先设计品牌企业大奖。

2016年 作品《大艺术家·源》荣获2016第四届ID+G金创意奖国际空间设计大赛陈列艺术类与地产设计类十佳设计奖；作品《大艺术家·形》荣获2016第四届ID+G金创意奖国际空间设计大赛陈列艺术类与地产设计类银奖；荣获2016第四届ID+G金创意奖国际空间设计大赛人物奖知名设计师奖；PINKI 品伊国际创意产业集团荣获第四届ID+G金创意奖国际空间设计大赛机构奖最具影响力设计机构；受聘为UIDM澳门国际设计联合会副会长；刘卫军总监与钟文萍总监受聘为广州大学松田学院艺术与传媒系创业导师；荣获2016祝融奖深圳赛区单项奖家居空间设计一等奖以及全国总评银奖，中国建筑装饰协会确定PINKI DESING作为中国建筑装饰行业职业教育实训基地，并授予刘卫军先生"中国建筑装饰行业职业教育特聘专家"称号；作品《大艺术家》广州南沙金茂湾项目荣获2016台湾TAKAO室内设计大赏金奖；作品《遇外时光》《厢》荣获2016台湾TAKAO室内设计大赏TAKAO优选奖；作品《大艺术家·源》《大艺术家·形》《Litang丽汤温泉公园生态酒店》《感·时光旅者》入围2016法国双面神INNODESIGN PRIZE国际设计大奖；作品《唯漫时光》《丽汤首山温泉公园生态酒店》分别荣获APDC赞颂·创造力2015/2016亚太室内设计精英邀请赛银奖和铜奖；2016先知·中国装饰设计行业大震荡暨中国设计大奖龍承奖深圳赛区启动盛典现场，前国务院副总理吴桂贤女士亲自为刘卫军先生颁发2016年度影响力人物奖；公司荣获第二届中国经济新模式创新与发展峰会暨2016中国行业领先品牌电视盛典2016中国最具商业价值十大设计机构、中国文化创意及室内空间环境艺术最具影响力十大领导品牌企业；刘卫军先生荣膺中国艺术设计2016十大首席人物；作品《大艺术家·厢》荣获2016美居奖中国最美样板间大奖。

LIST OF AWARDS

1998   Outstanding Designer and Youth Job Expert awarded by Shenzhen Construction Investment Holdings Group, the First Prize of Shenzhen Decoration and Design Competition;

1999   Shenzhen Model Project Award for the design of high technology exhibition hall of Shenzhen Technology Building;

2000   The Best Design Project Award given by the China Century Interior Design Forum;

2001   The Excellence Award and four nomination prizesof the China Interior Design Competition of CIID;

2002 The Excellence Award, the Third Prize and seven nomination prizes of the China Interior Design Competition of CIID; the First Prize and the Best Innovation Award of the Southern China Interior Design Competition; the gold award of the Asia Pacific Interior Design Awards (APIDA) which is the first important award he received;

2003 The Third Prize and eight nomination prizes of China Interior Design Competition of CIID on his third participation, APIDA Grand Award (the only awardee from club category in China regions);

2004 The Second Prize (project) of Mainland China, Taiwan, and Hong Kong Interior Design Competition; Shenzhen Construction Decoration Group Special Contribution Award; three prizes of China Interior Design Competition of CIID on his fourth participation, and the First Prize of National Commercial Space Design Competition;

2005 IAID the Most Influential Young Designer Award, two awards (projects and residential flats) of Shenzhen First Interior Design Festival; BDCI Most Popular Online Figure Award; Prize for Promotion of Interior Design in China for 20 Years; CIID Association Contribution Award;

2006 IDCFC City Outstanding Interior Designer Award, China's Top 10 Residential Designer; the Excellence Award of the China Interior Design Competition of CIID on his sixth participation; Nomination Award for Award of the Best Residential Space, Award of the Best Display, and the Best Office Space given by China (Shanghai) International Building andInterior Design Festival; 2004-2006 China's Best Interior Designers Honorary Title;

2007 "Man of Interior Design in China" of the Second (Boao) China Home Forum; the Best Conceptual Design Nomination Award of China (Shanghai) International Building and Interior Design Festival; Annual Outstanding Thesis Award for China's Interior Design for his thesis "How to balance environmental awareness" (《怎样平衡环保意识》), Prize for China Hotel Design Competition, Silver Award for Club Design in the 15th APIDA (the top award from club category in China regions), Annual Top Ten Cover Characters for Interior Design in China;

2008 Nomination Award for Best Decoration Award of China (Shanghai) International Building and Interior Design Festival, Innovative Design Award of the First Stone Material Design and Art Contest; the Excellence Award and the Best Work Award of the China Interior Design Competition of CIID on his seventh participation; one of the Top 10 Residence Interior Design Company in China; one of the Top 10 Interior Design Company of Greatest Value in China for PINKI DESIGN US IARI LiuWeiJun Design Firm;

2009 The Excellence Award for the Best Use of Colors of China (Shanghai) International Building and Interior Design Festival, Modern Decorations International Media Awards Annual Model Space Awards, Top Ten Most Influential Bullogs of SOHU, SOHU Second Designer Online

Media Award, Annual Outstanding Blogger Award. Modern Decorations Cup Shenzhen Designer Go-kart Competition the Most Powerful Team Award for the team under his leadership, Exceptionally Senior Researcher of China's Design Industry, National Design Industry Principal Expert, First Batch of National Outstanding Talent Model Medal of the Design Industry, PINKI DESIGN US IARI LiuWeiJun Design Firm was awarded 2009 Most Influential Design Institution of CIDF, Danfu Lau was admitted by CIDF as "Expert Committee Member", PINKI DESIGN US IARI LiuWeiJun Design Firm was awarded the Sliver Award of CIDF, News Figure of the Time in China, the "Excellence Award for Residential Space", "Third Prize for Club Space", "Second Prize for Model Suite", "Second Prize for Display Space" of China Interior Space Environmental Art Design Competition organized by China Building Decoration Association, Honorary Title of Experts of Sina Assessment Team; his work Navy Blue Residence (《藏青亭居》) was awarded China Lighting Design Competition Excellence Award of NVC Lighting Magazine Cup, Danfu Lau was awarded "Association Contribution Award" and "Outstanding Designer" Honorary Certificate on the 20th anniversary of CIID, Honorary Title of Cultural Development Outstanding Entity of Shenzhen Enterprises for PINKI DESIGN US IARI LiuWeiJun Design Firm;

2010 Listed as key nominee for annual influential figures for Harmonious China·2010; China Hotel Design Master conferred by the Fifth Local and Overseas Hotel Platinum Award; Jingtang Prize for Annual Outstanding Award for Yong He Fang (《永和坊》), Annual Club Space Award Finalists of the Eighth International Media Award of Modern Decorations for Knight Phantom (《爵士魅影》), 2010 Modern Decoration Eighth International Media Awards Annual Home Space Awards Finalist and International Space Design Awards Aite Award for Best Art Design Award for Cottage in the Dream (《梦遇香居》) and Honorary Title for "Most Influential Design Firm in Greater China for PINKI DESIGN;

2011 The 5th Cross-strait Interior Design Competition "Best Design Team Award", "Outstanding Construction Space Gold Award", "Outstanding Designer"; Sino Blog "2010 Active Award"; China (Shanghai) International Building and Interior Design Festival Golden Bund Award "Best Landscape Design Finalist Award" for Corner over the Clouds(《云顶天阁》); Honorary Title of "China's Top Ten High-end Residence Designer"; Quality Home (《美好家园》)China's 5th anniversary cum award presentation ceremony was held in Beijing Hong Kong Jockey Club Clubhouse, Danfu Lau was conferred "Interior Design Master Award" (other awardees are: T K Chu, Kenneth Ko, Ching Kao, Lai Yanan, Jamy Yang, Zi Xiangge, An Dong, Chen Yaoguang, the Design Republic, Design MVW, Jiang Qionger, Jeff Dayu Shi, Yang Ruidan, Ricky Wong); 6th CIDF "2010-2011 Outstanding Designer Award,CIID "Association Award Finalist Award for his works Poetry in the Woods (《丛林里的诗意漫游》) and Metropolitan Phantom (《都市魅影》); the Sixth Local and Overseas Hotel Platinum Award conferred the honor of "Top Ten Platinum Designer" to Danfu Lau, one gold award, two silver awards, one bronze award for Cross-strait Interior Design Competition, PINKI DESIGN 10th anniversary publication

Inspired Spatial Creation (《灵感塑造空间》) was conferred Jintang Award "Annual Best Publication Nomination Award", Jintang Award Annual Leisure Space Design Excellence Award for his work Sense - Time Traveler (《感·时光旅者》); 2011 Jintang Award: Annual Top Ten Model Suite Award for He Xu Yun Yi (《和煦蕴逸》);

2012 The 7th Golden Bund Award for Best Landscape Design for Sense - Time Traveler (《感·时光旅者》); China Display Art Development Forum "Annual Interior Display Art Pioneer Award; 2012 Jintang Award: Annual Top Ten Model Suite Award for Song of Flowers and Full Moon (《花好月圆曲》);

2013 8th IHFO 2013 Chinese and Foreign Hotel Forum Platinum Award for Litang Shoushan Hot Spring Park Hotel (《丽汤·温泉公园酒店》); 2013 CIDA China Interior Design Awards Annual Model Design Award for Baroque Love Song (《巴洛克恋曲》); 2013 CIDA China Interior Design Awards Annual Model Design Nomination Award for Leisure (《闲》); 2013 China Interior Design Grand Prix (CIID Society Award) Annual Program Bronze Award for Litang Shoushan Hot Spring Park Hotel (《丽汤·温泉公园酒店》); 2013 Jintang Award: Annual Top Ten Model Suite Award for Hao Sheng Yin Yi (《豪笙印溢》); 2013 Model Suite Design Excellence Award for a Slave of Love (《侬本多情》);

Meiju Award "2013 the Most Beautiful Space Award" for Yi Xin Yun Yi (《怡心韵驿》). Developing in the education sector, Danfu Lau was conferred "Contribution to Interior Design Education in China Award" in commemoration of the 25th anniversary of the China National Interior Decoration Association;

2014 The Creative Dimension of Inspiration book series were in the collection of the National Library; "Finalist Award" for Residential Projects of the 17th China Interior Design Competition for Carving Time (《雕刻时光》); 2014 Jintang Award "Annual Excellence Works Award" for Song - Style Residence (《曲悦·风尚居》); "Finalist Award" for Residential Projects of the 17th China Interior Design Competition in 2014 for Song - Style Residence (《曲悦·风尚居》); Annual Outstanding Designer Award of 12th Modern Decoration International Media Award in 2014; 10th Chinese Xinjiang International Art Biennale Exhibition "Excellence Award" for Song - Style Residence (《曲悦·风尚居》); 2014 Juran Cup CIDA China Interior Design Award "Living Space - Residence Design Award" for Song - Style Residence (《曲悦·风尚居》); First Prize of "China's Most Beautiful Model Suite" for North Competition Region of 2014 Meiju Award for Song - Style Residence (《曲悦·风尚居》); Danfu Lau was conferred "2014 Golden Phoenix Award for Tutor and China Soft Art" in "Xihu Soft Art Design Week" of "The East".

2015 "Special Contribution Award" of 2015 CWDA First International Display Window Design Competition Award Ceremony for "Daydreaming" (《白日梦》); 2015 China Interior Design Grand Prix (CIID Society Award) Bronze Award for Diffused Time (《唯漫时光》); 2015 Macau Design Exhibition "Golden Lotus" Cup International Design Master Invitational Silver Award for

Shoushan Hot Spring Park Hotel (《丽汤温泉公园酒店》); "2015 China Home Talent Designer of the Era" Honor;

2015 United Nations 70+ Chinese Contemporary Art and Creative Design Achievement Award for Oriental Charm, Clubhouse Design Work; 2015 Real Estate Design Meiju Award "China's Most Beautiful Model Suite Award" of "Carving Time Once Again" (《再续雕刻时光》); CPC Award 2015 Education Contributor Award; CPC Award 2015 Innovative Design Brand Enterprise Award for PINKI DESIGN.

2016 Great Artist – Source (《大艺术家·源》) was granted 2016ID+G "Golden Creative Award" International Space Design Awards Exhibition Art/Real Estate Space Ten Best Design Award; Great Artist - Form (《大艺术家·形》) was granted 2016ID + G "Golden Creative Award" International Space Design Awards Real Estate Space/Display Art Silver Award; 2016ID + G "Golden Creative Award" International Space Design Awards Talent Award: Famous Designer Award; PINKI INTERNATIONAL CREATIVE INDUSTRY GROUP was awarded the 4th"Golden Creative Award" International Space Design Awards: Most Influential Design Firm; appointed as UIDM Vice-chairman; Danfu Lau and Zhong Wenping were engaged by Guangzhou University Sontan College as Creative Tutor for the Faculty of Arts and Media; 2016 Vulcan Award of Shenzhen Competition Region of Single Award Residential Space Design First Prize and National Overall Assessment Silver Award; CBDA designated PINKI DESIGN as the vocational education training base for China's building decoration and conferred the title of China's Building Decoration Industry Vocational Training Special Expert; Zhong Wenping was appointed by Guangzhou University Sontan College as entrepreneurial tutor for the Faculty of Arts and Media; 2016 Taiwan TAKAO Interior Design Award Gold Award for Great Artist (《大艺术家》), Guangzhou Nansha Jin Mao Harbor Project; Taiwan TAKAO Interior Design Award TAKAO Merit Award for "External Time" (《遇外时光》), "Apartment" (《厢》); 2016 INNODESIGN PRIZE finalists for Great Artist – Source (《大艺术家·源》); Great Artist – Form (《大艺术家·形》); Litang Shoushan Hot Spring Park Hotel (《Litang丽汤温泉公园生态酒店》), Sense – Time Traveler (《感·时光旅者》); Silver and Bronze Awards for APDC Celebrating Creativity 15/16 Asia Pacific Interior Design Awards for Elite for Diffused Time (《唯漫时光》) and Litang Shoushan Hot Spring Park Hotel (《Litang丽汤温泉公园生态酒店》); 2016 Influential Person Award presented by Wu Guixian, former vice premier of the State Council at 2016 Prophecy-China Decoration Design Industry Brainstorming cum China Design Award Long Cheng Award Shenzhen Competition Region Launching Ceremony; the company was awarded 2016 China's Top Ten Design Firms of Greatest Commercial Value, Chinese Cultural Innovation and at the Second Summit of New Mode of Innovation of Chinese Economy cum 2016 China's Leading Brand Television Ceremony, Most Influential Top Ten Leading Branded Enterprise for Chinese Cultural Innovation and Interior Spatial Environment; Danfu Lau won China Arts and Design – 2016 Top Ten Persons; 2016 Meiju Award China's Most Beautiful Model Suite Award for the Great Artist – Apartment.

客户列表
LIST OF CLIENTS

[TOP PROPERTY DESIGN BRANDFOR 17YEARS] THANK YOU FOR BEING WITH US...

17年顶级地产设计知名品牌
感谢一路有您……

CHINA JINMAO HOLDINGS GROUP LIMITED
中国金茂集团

世界500强企业之一中国中化集团公司（2016年列《财富》全球500强之第139位）旗下房地产和酒店板块的平台企业
A platform enterprise under the real estate and hotel segments of Sinochem Group, one Fortune 500 enterprises (ranking the 139th among the Fortune 500 enterprises in 2016)

POLY REAL ESTATE
保利地产

中国房地产行业领导公司品牌
China's leading company brand in the real estate industry

CHINA WANDA REAL ESTATE
中国万达集团

中国商业地产领导者
Leader in China's commercial property

VANKE PROPERTY
万科地产

全球最大的专业住宅开发商
The world's largest professional residential property developer

---

EVERGRANDE GROUP
恒大集团

世界500强企业集团中国标准化运营精品地产领导者
Fortune 500 enterprise China's boutique property leader in standardization operations

---

OCT PROPERTIES
华侨城地产

中国旅游地产领导者
China's tourism property leader

CHINA RESOURCES LAND
华润地产

全球500强企业之一，《财富》杂志发布的2015年世界500强排行榜中排名第115位
One of the Fortune 500 enterprises (ranking 115th among the Fortune 500 enterprises in 2015)

CHINA OVERSEAS PROPERTY
中海地产

连续13年入选"中国蓝筹地产"企业，并荣获"中国价值地产总评榜·年度企业公民
Selected as "China Real Estate Blue Chip" for 13 consecutive years, and was awarded "China Valuable Real Estate Awards – Corporate Citizen of the Year"

YiDA
亿达集团

中国领先的商务园区运营商
China's leading commercial park operator

DAYU GROUP
大禹集团

中国名盘50强
China Top-50 Real Estate Projects

AVIC REAL ESTATE
中航地产

连续6年入围深圳百强企业
Top-100 enterprises in Shenzhen for six consecutive years

BRC
蓝光集团

连续6年入围深圳百强企业
Top-100 enterprises in Shenzhen for six consecutive years

ZHONGLIANG REAL ESTATE GROUP
中梁地产

中国房地产百强企业，2017年跻身综合实力40强，成长性TOP10，融资能力TOP10
Top-100 real estate developers in China, one of the 40 most competitive Chinese developers, top-10 fast-growing real estate companies and top-10 financing ability in 2017

HUAXIA REAL ESTETE
华夏地产

国家房地产开发一级资质的外资企业
Foreign-funded enterprise with Grade-1real estate development qualifications awarded by the state

---

GEMDALE CORPORATION
金地集团

连续13年蝉联"中国蓝筹地产"，连续14年跻身"沪深房地产上市公司综合实力10强"，连续5年位列"中国责任地产TOP10"
China Real-Estate Blue Chip for 13 consecutive years, Shenzhen Top10 Listed Real Estate Companies for 14 consecutive years and Top10 China Responsible Real Estate Enterprises for 5 consecutive years

---

GREENTOWN REAL ESTATE
绿城地产

中国理想生活综合服务商第一品牌
No. 1 brand of integrated service provider for ideal lifestyle in China

---

SUMITOMO REALTY & DEVELOPMENT CO.,LTD
日本住友不动产株式会社

《福布斯》2014年全球企业2000强中排名第588位
In "Forbes" in 2014 the world's top 2000 companies ranked in the top 588th.

PINKI DSIGN

创新　创意　创造价值

用 设 计 复 兴 人 文 美 学 经 典

快 乐 生 活　　快 乐 工 作

ENJOY LIFE, ENJOY DESIGN